α, β, γ . . . Z

A Primer in Particle Physics

$\alpha, \beta, \gamma \ldots Z$
A Primer in Particle Physics

By L.B. Okun
Institute of Theoretical and Experimental Physics,
Moscow, USSR

Translated from the Russian by V.I. Kisin

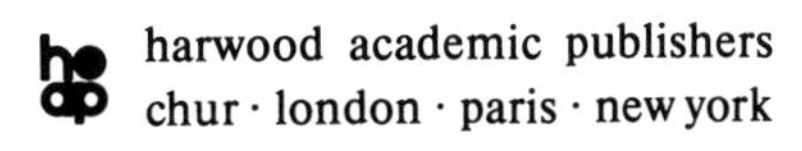
harwood academic publishers
chur · london · paris · new york

Harwood Academic Publishers

Post Office Box 197
London WC2E 9PX
England

58, rue Lhomond
75005 Paris
France

Post Office Box 786
Cooper Station
New York, New York 10276
United States of America

Library of Congress Cataloging-in-Publication Data

Okuń, L.B. (Lev Borisovich)
[Alpha], [beta], [gamma] — [zeta].
Includes index.
1. Particles (Nuclear physics) I. Title.
II. Title: A primer in particle physics.
QC793.2.036 1987 539.7′21 86-22906
ISBN: 3-7186-0374-8 (Hardback)
ISBN: 3-7186-0405-1 (Paperback)

 Printed in Great Britain

Dedicated to the memory of
Isaak Yakovlevich Pomeranchuk

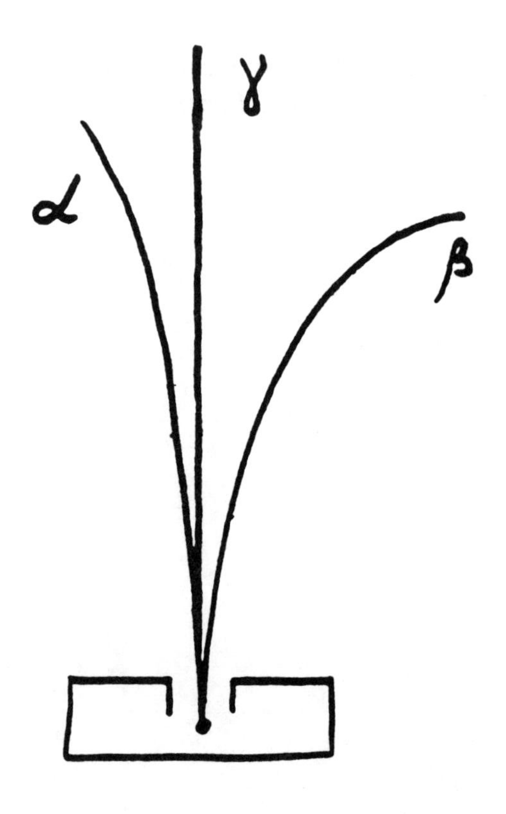
γ
α
β

Contents

$\alpha, \beta, \gamma \ldots Z$

A Primer in Particle Physics

by L.B. Okun

ERRATA

Page 3, line 23: 10^{-16} cm should read 10^{-18} m

Page 4, line 3: 0.01% should read 1%

Page 11, line 29 should read: When $s > 0$, the interval is said to be timelike; when $s < 0$, it is spacelike; when $s = 0$, it is lightlike.

Page 20, line 7: $e^4 m_e/2\hbar$ should read $e^4 m_e/2\hbar^2$

Page 22, line 3: $\lambda = \omega c$ should read $\lambda\omega = c$

Page 35, line 22 should read: The spin **J** of a meson is the vector sum ...

Page 36, line 5 should read: There are numerous mesons for which $L \neq 0$ and $J > 1$ ...

Page 36, line 10 should read: and the spin **J** of the nucleon equals the vector sum ...

Page 36, lines 17–20: MэB should read MeV

Page 46, line 3: $\tau^+ \to \tilde{\nu} e^+ \nu_e$ should read $\tau^+ \to \tilde{\nu}_\tau e^+ \nu_e$

Page 63, line 7 and figure caption: $\tilde{\nu}_e e$ should read $\bar{\nu}_e e$

Page 66, line 20 should read: Only the weak interaction is powerful enough ...

Preface

The subject of this book is the physics of elementary particles and the fundamental forces between them. I shall begin with a brief comment about its title.

Modern studies of the fundamental forces between particles began with the discovery of radioactivity in 1896 and was followed by investigations into the properties of α, β and γ rays. The long-awaited (but nevertheless sensational) discovery of the W and Z bosons in 1983 concluded a long period of research. Hence, the title 'α, β, γ . . . Z'.

Nevertheless, this is not a book about the history of physics. It describes its present state and possible future development. In fact, the discovery of the W and Z bosons marks the beginning of a new and very promising journey. Physics is not an alphabet, and Z is not the end of it, though in a sense, the title α, β, γ . . . Z hints that this book is a sort of A-to-Z of the foundations of modern physics.

The book grew out of popular physics lectures that I was invited to deliver time and again to audiences who knew very little about particle physics, or even about physics in general. The last of these lectures was given in the summer of 1983, soon after the discovery of the Z boson. The idea of writing this book took shape when I was thinking about questions asked during this past lecture.

My aim was to write a book that could be read and understood by students who were about to graduate, or have graduated, from high school and who were actively interested in physics. I hope that my prospective reader leafs more or less regularly throught popular science periodicals and has read at least some of the more elementary physics books.

The most serious pitfall that I was trying to avoid on every page was an involuntary tendency to share with the reader not only the basic physics but also the less important details that delight the specialist so much but so often frustrate the beginner. I am afraid that while additional 'weeding' might be welcome in some places, overweeding may be evident elsewhere.

The selection of the basic information, with everything else ruthlessly dropped, proved to be an interesting task. At first I wanted to employ the bare minimum of special terms and concepts. But as the writing progressed, I discovered that some phenomena could not be explained without certain concepts that I originally intended to avoid, and the text became more and more complicated toward the end of the book. Indeed, one of the main difficulties encountered in entering any new branch of science is a flood of new words. Even if each one of them has a simple meaning, keeping them all in mind is not at all easy for the uninitiated. To help the reader, the Subject Index at the end of the book provides a sort of crib, i.e. a list of the basic concepts of elementary-particle physics.

The physics of elementary particles is often referred to as high-energy physics. Even a cursory glance reveals that the processes studied in high-energy physics are very unusual, and their exotic properties challenge the imagination.

Eventually, we come to realize that, in many respects, these processes differ from ordinary phenomena such as, say, the burning of wood, not qualitatively but only quantitatively, e.g. by the amount of of energy released. I therefore begin with a brief discussion of such seemingly familiar concepts as *mass, energy* and *momentum*. The correct understanding of these three terms will help us with the following pages.

The key concept in fundamental physics is the idea of a *field*. I begin with examples that are familiar from school, and then gradually introduce the multitude of fascinating properties of quantum fields. I have made a deliberate attempt to provide a simple explanation of things that allow a more or less simple explanation. But I must emphasize that quite a few things in modern physics defy elementary interpretation, and the reader will be able to master such ideas only after a good deal of serious work with other, more complicated books.

The draft text of the book was completed in October of 1983. It was read by L.G. Aslamazov, V.I. Kisin, A.V. Kogan, V.I. Kogan, A.B. Migdal, B.L. Okun, Ya.A. Smorodinskii, and Ya.B. Zel'dovich, who made very useful suggestions that enabled me to simplify the original text, sometimes dropping a number of relatively complicated passages and sometimes adding more detailed explanations. I am grateful to them for their advice. Useful suggestions were proposed by Alan Kostelecký, who read the English translation of the book. I am also grateful to E.G. Gulyaeva and I.A. Terekhova for their help in preparing the manuscript for publication.

I would like to express my gratitude to Carlo Rubbia for permission to reproduce the schematic drawings of the detector in which intermediate bosons were discovered.

I have special feelings of affection and gratitude for my teacher Isaak Yakovlevich Pomeranchuk, who introduced me to the world of elementary particles and taught me my profession. Academician I.Ya. Pomeranchuk lived a short life (1913–1966) but left behind him the image of a man passionately and unselfishly devoted to science. He had a keen interest in everything new and was mercilessly critical and self-critical as a scientist for whom the achievements of his colleagues were a source of great happiness. This image of him is still fresh in the minds of those who knew him. I dedicate this book to his memory.

L.B. Okun

Particles and interactions: a minidigest

Atoms consist of *electrons* e, forming electron shells, and *nuclei*. Nuclei consist of *protons* p and neutrons n. Protons and neutrons consist of two types of *quark*, u and d: p = uud, n = ddu. A free neutron undergoes *beta-decay*: n $\rightarrow$ pe$\tilde{\nu}$ where $\tilde{\nu}_e$ is the *electron antineutrino*. The decay of the neutron is caused by the decay of the d quark: d $\rightarrow$ ue$\tilde{\nu}_e$.

The attraction between an electron and a nucleus is an example of the *electromagnetic interaction*. The attraction between quarks is an example of the *strong interaction*, whilst beta decay is an example of the *weak interaction*. In addition to these three *fundamental interactions* there is a fourth important fundamental interaction, namely, the *gravitational* interaction, which makes all particles attract one another.

Fundamental interactions are described by the appropriate *fields of force*. The *excitations* (the *quanta*) of these fields are the particles called *fundamental bosons*. The quantum of the electromagnetic field is the *photon*, the quanta of the strong field are the eight *gluons*, the quanta of the weak field are the three *intermediate bosons* W^+, W^-, Z^o, and the quantum of the gravitational field is the *graviton*. All fundamental bosons, except the graviton, have been observed experimentally.

Most particles are classified into pairs of "twins", i.e. particle-antiparticle pairs. An *antiparticle* has the same mass as the particle but opposite charge (electric charge, weak charge, and so on). Particles identical to their antiparticles, i.e. those with no charge, are said to be *truly neutral*. An example is the photon.

In addition to the electron e and the electron neutrino ν_e, we know two more pairs of particles similar to them; μ, ν_μ and τ, ν_τ. All six are collectively known as *leptons*. In addition to the u and d quarks, there are *two other pairs of more massive quarks*: c, s and t, b (the existence of the t-quark still lacks experimental confirmation). The leptons and the quarks are called *fundamental fermions*.

Particles consisting of three quarks are called *baryons*, and those consisting of a quark and an antiquark are called *mesons*. Baryons and mesons taken together form the family of strongly interacting particles, called *hadrons*.

Basic particles: the electron, proton, neutron, and photon

Elementary-particle physics deals with those tiniest particles that are the building blocks of the world around us and of ourselves. Our aim in this book will be to examine the internal structure of the particles, to analyze the processes in which they participate, and to establish the laws that these processes obey.

The basic experimental technique (although not the only one!) in elementary-particle physics is to make a beam of high-energy particles collide with a stationary target or with another beam. The higher the energy of collision, the richer the interactions between the particles and the larger the amount of information about them that we can extract. It is for this reason that elementary-particle physics and high-energy physics have become almost synonymous. However, in this book, our acquaintance with particles begins not with high-energy collisions but with ordinary atoms.

The fact that matter consists of atoms with diameters of about 10^{-10}m is common knowledge. The atomic diameter is the size of the electron shell of the atom. However, practically the entire mass of the atom is confined to its nucleus. The nucleus of the hydrogen atom (the lightest atom) consists of a single proton, and its atomic shell consists of a single electron. We note, in passing, that one gram of hydrogen contains $6 \cdot 10^{23}$ atoms, so that one proton mass is approximately $1.7 \cdot 10^{-27}$ kg (the electron mass is smaller by a factor of roughly 2000). The nuclei of heavier atoms contain neutrons as well as protons. The symbols are: e for the electron, p for the proton, and n for the neutron.

The number of protons in each atom is equal to the number of electrons. The electric charge of a proton is positive and that of an electron is negative, which means that each atom is electrically neutral. Atoms of a given chemical element whose nuclei contain the same number of protons but a different number of neutrons are called its *isotopes*. For example, besides ordinary hydrogen we know two heavier hydrogen isotopes, namely, deuterium and tritium, whose nuclei contain one and two neutrons, respectively. These three hydrogen isotopes are denoted by ^1H, ^2H, ^3H, where the superscript represents the total number of protons and neutrons in the nucleus. The nucleus of deuterium is referred to as deuteron, which we will represent by D (the letter d is sometimes used instead in the scientific literature).

Ordinary hydrogen, ^1H, is the most abundant element in our Universe. The second in abundance is the helium isotope ^4He, whose electron shell contains two electrons and whose nucleus contains two protons and two neutrons. Ever since radioactivity was discovered, the nucleus of the isotope ^4He has had a special name — the α particle. The helium isotope ^3He, whose nucleus consists of two protons but only one neutron, is less abundant.

The proton and neutron have almost equal radii (about 10^{-15} m). The masses of these particles are also nearly equal: the neutron is a mere 0.1% heavier than the proton. Neutrons and protons are rather closely packed in atomic nuclei, so that the volume of the nucleus is approximately equal to the sum of volumes of the constituent *nucleons* (''nucleon'' stands for either ''neutron'' or ''proton'', and is used when differences between the two are unimportant).

As for the size of the electron, it is too small to measure with existing techniques, and we know only that the radius of the electron is certainly less than 10^{-16} cm. For this reason, electrons are usually said to be pointlike particles.

The electrons in an atom are sometimes compared with the planets of the solar system. This analogy is false in many respects. First, the motion of an electron differs qualitatively from the motion of a planet because electrons do not obey the laws of *classical* mechanics. They follow the laws of *quantum* mechanics, which will be discussed later. Here we merely note that, because of the quantum-mechanical nature of the electron, an ''instantaneous photograph'' of the atom would reveal that the electron has an appreciable probability of being found at any instant at any point of its orbit (and even outside the orbit). In contrast, the position of a planet in its orbit is dictated by the laws of classical mechanics and can be predicted with very high accuracy. If we compare a planet with a streetcar on rails, the electron behaves more like a taxicab.

A number of purely quantitative differences that destroy any similarity between atomic electrons and planets must also be mentioned at this point. For example, the ratio of the radius of the electron orbit in an atom to the radius of the electron itself is much greater than the ratio of the radius of the Earth's orbit to the radius of the globe. The electron in a hydrogen atom travels with a velocity of about 0.01% of the velocity of light[†] and completes about 10^{16} orbits in one second. This is approximately a million times greater than the number of orbits completed by the Earth around the Sun in the entire lifetime of the solar system. Electrons in the inner shells of heavy atoms move even faster: their velocities reach two-thirds of the velocity of light.

The velocity of light in vacuum is usually denoted by the letter c. This fundamental physical quantity has been measured to a very high accuracy. Its value is

$$c = 2.997\,924\,58\,(1.2) \cdot 10^8 \text{ m/s.}^{\ddagger}$$

An approximate value is: c = 300 000 km/s.

This is a good place to introduce the particle of light, i.e., the *photon*. Photons are not the components of atoms in the way that electrons and nucleons are. It is therefore common to speak of photons not as particles of matter but as particles of radiation. But in fact the role of photons in the machinery of the Universe is no less important than that of electrons and nucleons.

The energy of a photon determines how it is to be classified: we speak of radio waves, infrared radiation, visible light, ultraviolet radiation, x-rays, and the so-called nuclear γ-ray quanta. The higher the energy of these quanta, the more penetrating (or "harder") they are. Hard quanta are capable of passing through quite thick metal shields. In elementary-particle physics, the symbol for photons is γ, regardless of their energy.

The principal feature that distinguishes photons from all other particles is that they are very readily created and just as readily annihilated. Strike a match, and you create billions of photons. A piece of black paper placed in the way of a beam of visible light is sufficient to absorb the photons. Of course, the efficiency with which a screen absorbs, transforms, and re-emits incident photons depends both on the nature of the screen and on the photon energy. X-rays and hard γ-rays are not absorbed as readily as

---

[†]More accurately, the ratio of the velocity of the electron in the hydrogen atom to the velocity of light is approximately 1/137. Try not to forget this number. You will encounter it more than once on the pages of this book.

[‡]Throughout this book, the figures in parentheses (in similar cases) give the experimental uncertainty in the last digits of the main number.

visible light. At very high energies, the difference between photons and other particles is hardly greater than the differences between the different species of these particles. At any rate, it is far from simple to create or absorb high-energy photons. However, the lower the photons' energy, the easier it is to create and absorb them.

One of the spectacular attributes of photons, which to a great extent determines their remarkable properties, is that their mass is zero. When a particle has nonzero mass, we know that lower energy corresponds to slower motion. A massive particle is allowed not to move at all, i.e., to be at rest. However, a photon always propagates with the velocity of light, whatever its energy.

Mass, energy, momentum, and angular momentum in Newtonian mechanics

We have already used the words "energy" and "mass" several times. Let me now explain what they mean. I shall also discuss *momentum* and *angular momentum*. All these physical quantities (mass, energy, momentum, and angular momentum) play a fundamental role in physics. Their importance stems from the fact that, in an isolated system of particles, the total energy and momentum of the system, as well as its total angular momentum and mass, are conserved, i.e., do not change with time regardless of the complexities of interactions between the constituent particles.

We begin our discussion with Newtonian mechanics. Consider a body of mass m and velocity **v**.† In Newtonian mechanics, the momentum of a body is

$$\mathbf{p} = m\mathbf{v}$$

and its *kinetic energy* is

$$T = \tfrac{1}{2} m\mathbf{v}^2 = \mathbf{p}^2/2m$$

where $\mathbf{v}^2 = v_x^2 + v_y^2 + v_z^2$ and v_x, v_y, v_z are the projections (components) of the vector **v** on the coordinate axes.

---

†Here and throughout the book, boldface letters represent vectors, i.e., quantities that have both magnitude and direction in space.

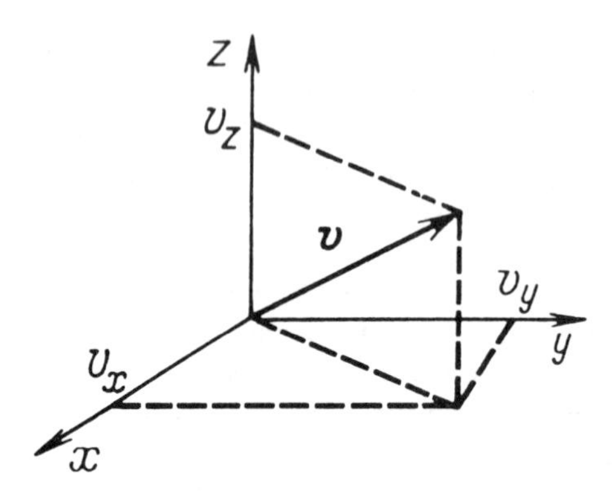

Figure 1. Projections of velocity vector **v** on the coordinate axes.

The coordinate system can be rotated in space in an arbitrary manner without affecting the value of $\mathbf{v}^2$. However, the directions and the magnitudes of the vectors **v** and **p** do depend on the magnitude and direction of the velocity of the coordinate system in which the motion of the body is described: they are said to depend on the *frame of reference*. For example, your house is at rest in a reference frame fixed to the Earth. In a reference frame fixed to the Sun, it moves with the speed of 30 km/s.

The quantity called angular momentum plays an important role in the description of rotational motion. As an example, let us examine the simple case of a particle (a mass point) moving on a circular orbit of radius $r = |\mathbf{r}|$ with constant speed $v = |\mathbf{v}|$ (r and v are the magnitudes, i.e., absolute values, of the vectors **r** and **v**, respectively). The angular momentum **L** of the orbital motion is then defined as the vector product of the position vector **r** and the momentum **p** of the particle, i.e., $\mathbf{L} = \mathbf{r} \times \mathbf{p}$.

Although the directions of the two vectors **r** and **p** vary with time, the direction of **L** remains unchanged. This is readily verified by inspecting Figure 2.

■ By definition, the vector product $\mathbf{a} \times \mathbf{b}$ of two vectors **a** and **b** is the vector **c** whose magnitude is

$$|\mathbf{c}| = |\mathbf{a}|\ |\mathbf{b}|\ \sin\theta,$$

where θ is the angle between **a** and **b**. The vector **c** is perpendicular to the plane containing the vectors **a** and **b**, so that **a**, **b**, and **c** form a so-called right-handed triple (in accordance with the familiar Ampère, or corkscrew, rule):

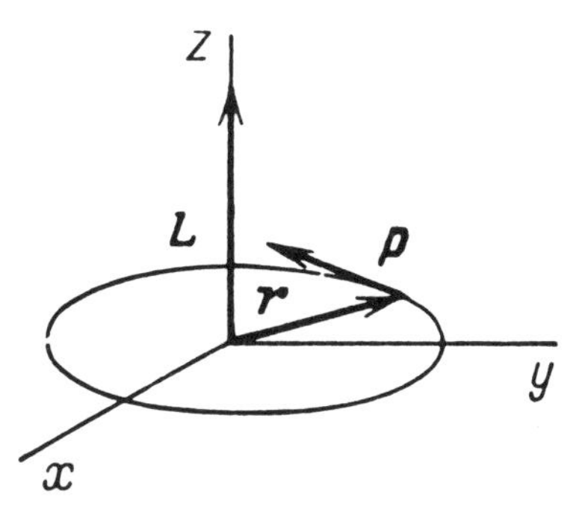

Figure 2. Orbital angular momentum **L** of a particle with momentum **p** moving in a circular orbit of radius **r**.

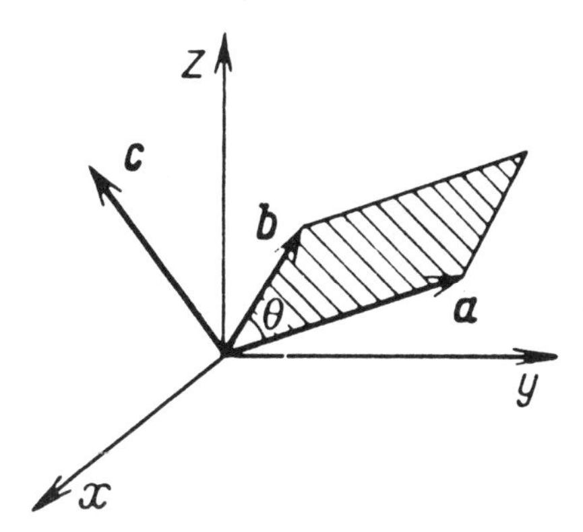

Figure 3. Vector **c** is the vector product of vectors **a** and **b**.

The components of the vector product are

$$c_x = a_y b_z - a_z b_y,$$
$$c_y = a_z b_x - a_x b_z,$$
$$c_z = a_x b_y - a_y b_x.$$

It is readily verified that, in the case shown in Figure 2,

$$L_x = L_y = 0, \; L_z = |\mathbf{r}| \; |\mathbf{p}| = \text{const.}$$

Having defined the *vector* product, we must also mention the *scalar* product of two vectors **a** and **b**, denoted by $\mathbf{a} \cdot \mathbf{b}$. By definition,

$$\mathbf{a}\cdot\mathbf{b} = a_x b_x + a_y b_y + a_z b_z.$$

You can easily check (see Figure 3) that $\mathbf{a}\cdot\mathbf{b} = |\mathbf{a}|\ |\mathbf{b}| \cos\theta$ and that arbitrary rotation of the mutually orthogonal (Cartesian) axes x, y, z does not affect the scalar product.

Note that three mutually orthogonal unit vectors pointing along the three axes are called *basis vectors*. They are usually denoted by $\mathbf{n}_x$, $\mathbf{n}_y$, $\mathbf{n}_z$ (Figure 4). It is clear from the definition of the scalar product that $a_x = \mathbf{a}\cdot\mathbf{n}_x$. ■

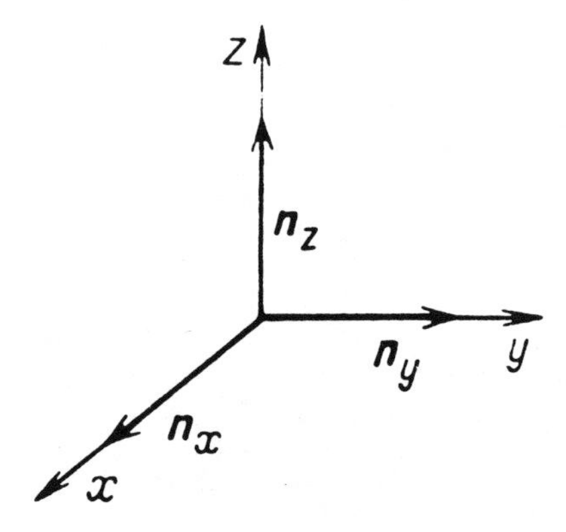

Figure 4. Three mutually orthogonal unit vectors.

The orbits of planets in the solar system are elliptical and not circular, which means that the separation between a planet and the Sun is a periodic function of time. The magnitude of the velocity of a planet also varies periodically. However, its orbital angular momentum remains constant (as an exercise, try deriving from this Kepler's second law, which states that the position vector of a planet sweeps out equal areas in equal times).

In addition to the orbital angular momentum characterizing its motion around the Sun, the Earth possesses (like all planets) an intrinsic angular momentum that characterizes its diurnal rotation. The conservation of the intrinsic angular momentum explains the behaviour of gyroscopes. The intrinsic angular momentum of an elementary particle is called its *spin*.

Mass, energy, and momentum in Einstein's mechanics

Newtonian mechanics provides a perfect description of the motion of a body when its velocity v is much less than the velocity of light: $v \ll c$. But this theory is manifestly incorrect when the velocity is comparable with c, especially when $v = c$. To describe motion with arbitrary velocity, up to that of light, we have to turn to Einstein's special theory of relativity, i.e., relativistic mechanics. Newton's non-relativistic mechanics is but a particular (although very important) limiting case of Einstein's relativistic mechanics.

The word "relativity" originates in Galileo's relativity principle. In one of his books, Galileo explained in very graphic terms that no mechanical experiment performed inside a ship could establish whether the ship is at rest or moving relative to the shore. Of course, this would be easy to do by simply looking out. But if one were in a cabin without portholes, then rectilinear motion of the ship could not be detected.

Galileo's relativity can be shown mathematically to demand that the equations of motion of bodies, i.e., the equations of mechanics, must be identical in all the so-called inertial reference frames. These are coordinate systems attached to bodies moving uniformly along straight lines relative to very distant stars. (Obviously, in the case of Galileo's ship, we disregard the diurnal rotation of the Earth, its rotation around the Sun, and the rotation of the Sun around the center of our Galaxy.)

Einstein's extremely important achievement was that he extended Galileo's relativity principle to all physical phenomena, including electrical and optical processes involving photons. This generalization of Galileo's principle has resulted in drastic changes in fundamental ideas, such as those of space, time, mass, momentum, and energy. For example, in Einstein's relativity the concepts of total energy and rest energy are introduced. The kinetic energy T is related to total energy E by

$$E = E_0 + T$$

where E_0 is the rest energy, related to the mass m of a body by the famous formula

$$E_0 = mc^2.$$

The mass of photon being zero, its rest energy E_0 is also zero. For a photon, "a rest is but a dream": it always moves with the velocity of light

c. Other particles, e.g., electrons and nucleons, have nonzero masses and therefore nonzero rest energy.

The relations between energy, velocity, and momentum for particles with $m \neq 0$ take the following forms in Einstein's mechanics:

$$E = \frac{mc^2}{\sqrt{(1-v^2/c^2)}}, \quad \mathbf{p} = E\mathbf{v}/c^2.$$

Consequently,

$$m^2c^4 = E^2 - \mathbf{p}^2c^2.$$

Each of the two terms on the right-hand side of the latter equation increases as the body travels faster, but the difference remains constant (physicists prefer to say that it is invariant). The mass of a body is a relativistic invariant, since it is independent of the reference frame in which the motion of the body is described.

Einstein's relativistic formulas for momentum and energy become identical with the non-relativistic Newtonian expressions when $v/c \ll 1$. Indeed, if we expand the right-hand side of the relation $E = mc^2/\sqrt{1-v^2/c^2}$ in a series in powers of the small parameter v^2/c^2, we obtain the expression

$$E = mc^2 \left[1 + \frac{1}{2}\frac{v^2}{c^2} + \frac{3}{8}\left(\frac{v^2}{c^2}\right)^2 + \ldots\right],$$

where the dots represent higher powers of v^2/c^2.

■ When $x \ll 1$, a function $f(x)$ can be expanded in a series in powers of the small parameter x. By differentiating the left- and right-hand sides of the expression

$$f(x) = f(0) + xf'(0) + \frac{x^2}{2!}f''(0) + \frac{x^3}{3!}f'''(0) + \ldots,$$

and each time evaluating the result for $x = 0$, you will readily verify its validity (when $x \ll 1$, the discarded terms are negligible). Thus,

$$f(x) = (1 - x)^{-1/2}, \; f(0) = 1,$$
$$f'(x) = \tfrac{1}{2}(1 - x)^{-3/2}, \; f'(0) = \tfrac{1}{2},$$
$$f''(x) = \tfrac{3}{4}(1 - x)^{-5/2}, \; f''(0) = \tfrac{3}{4}. ■$$

Note that for the Earth moving in its orbit with a velocity of 30 km/s, the ratio v^2/c^2 is 10^{-8}. For an air-liner flying at 1000 km/h, this parameter is much smaller: $v^2/c^2 \approx 10^{-12}$. Consequently, the nonrelativistic relations

$$T = \frac{1}{2} mv^2, \mathbf{p} = m\mathbf{v}$$

are valid for the airliner to an accuracy of about 10^{-12}, and the relativistic corrections are definitely negligible.

Let us now return to the formula that relates mass, energy, and momentum, rewriting it in the form

$$m^2c^2 = (E/c)^2 - p_x^2 - p_y^2 - p_z^2.$$

The fact that the left-hand side of this equality is unaffected by going from one inertial reference frame to another is similar to the invariance of the square of momentum

$$\mathbf{p}^2 = p_x^2 + p_y^2 + p_z^2,$$

or of any squared three-dimensional vector, under rotations of the coordinate system (see above, Figure 1) in ordinary Euclidean space. In terms of this analogy, m^2c^2 is said to be the square of a four-dimensional vector, namely the four-dimensional momentum p_μ, where the subscript μ takes on the four values $\mu = 0, 1, 2, 3$ and $p_0 = E/c$, $p_1 = p_x$, $p_2 = p_y$, $p_3 = p_z$. The space in which the vector $p_\mu = (p_0, \mathbf{p})$ has been defined is said to be pseudo-Euclidean. The prefix "pseudo" signifies in this case that the invariant is not the sum of the squares of all four components but is the expression

$$p_0^2 - p_1^2 - p_2^2 - p_3^2.$$

The transformation that relates the time and space coordinates in two different inertial systems is called the Lorentz transformation. Without writing out this transformation in full, we note that if two events are separated by t in time and $\mathbf{r}$ in space, the quantity

$$s = (ct)^2 - \mathbf{r}^2,$$

called the *interval*, remains constant under the Lorentz transformation, i.e., it is a Lorentz invariant. Note that neither t nor $\mathbf{r}$ are invariants individually. When $s < 0$, the interval is said to be timelike; when $s > 0$, it is spacelike; when $s = 0$, it is lightlike. When $s < 0$, two spatially distinct events can be simultaneous in one reference frame but nonsimultaneous in another.

Let us now consider a system of n noninteracting free particles. Let E_i be the energy of the i-th particle, $\mathbf{p}_i$ be its momentum, and m_i its mass.

The total energy and momentum of the system are, respectively,

$$E = \sum_{i=1}^{n} E_i \text{ and } \mathbf{p} = \sum_{i=1}^{n} \mathbf{p}_i.$$

Since, by definition, the mass is given by

$$M^2 = E^2/c^4 - \mathbf{p}^2/c^2,$$

the mass of the system is not, in general, equal to the sum of the masses of its constituents.

In our daily life, we are used to the equality $M = \sum_{i=1}^{n} m_i$, but this fails for fast particles. For example, the combined mass of two electrons colliding head-on with equal energies is $2E/c^2$, where E is the energy of each electron. In accelerator experiments, this combined mass is greater by many orders of magnitude than the quantity $2m_e$, where m_e is the mass of the electron.

We conclude this section with a few remarks about terminology.

Some books and popular-science periodicals use the phrases "rest mass" and "motional mass" (or the equivalent phrase "relativistic mass"). The latter mass increases with increasing velocity of the body. The "rest mass" m_0 is then taken to mean the physical quantity that we have simply called *mass* and denoted m, and the *relativistic mass* m is taken to mean the energy of the body divided by the square of the velocity of light, $m = E/c^2$, which definitely increases with increasing velocity. This obsolete and essentially inadequate terminology was widespread at the beginning of this century when it seemed desirable, for purely psychological reasons, to retain the Newtonian relation between momentum, mass, and velocity: $\mathbf{p} = m\mathbf{v}$. However, as we approach the end of the century, this terminology has become archaic and only obscures the meaning of relativistic mechanics for those who have not mastered its foundations sufficiently well.

In relativistic mechanics, the "rest mass" is neither the inertial mass (i.e., the proportionality coefficient between force and acceleration) nor the gravitational mass (i.e., the proportionality coefficient between the gravitational field and the gravitational force acting on a body). It must be emphasized that the infelicitous "motional mass" cannot be interpreted in this way, either.

The correct relation between the force **F** and the acceleration $d\mathbf{v}/dt$ can be found if we use the relativistic expression for the momentum

$$\mathbf{p} = m\mathbf{v}/\sqrt{(1-v^2/c^2)},$$

and recall that $\mathbf{F} = d\mathbf{p}/dt$. The formula $\mathbf{F} = m\mathbf{a}$, which is familiar from school textbooks, is valid only in the non-relativistic limit in which $v/c \ll 1$.

That the gravitational attraction is not determined by the "rest mass" is evident, for example, from the fact that the photon is deflected by a gravitational field despite its zero "rest mass". The gravitational attraction exerted on the photon by, say, the Sun, increases with increasing photon energy. We are therefore tempted to say that the phrase "motional mass" is meaningful at least in this case. In fact, this is not so. A consistent theory of the motion of a photon (or any object moving with velocity comparable with the velocity of light) in a gravitational field will show that the energy of a body is *not* equivalent to its gravitational mass.

To conclude this discussion of "mass", I must ask the reader never to use the phrases "rest mass" or "motional mass" and always to mean by "mass" the relativistically invariant mass of Einstein's mechanics.

Another example of unfortunate terminology is the false claim that high-energy physics and nuclear physics are somehow able to transform energy into matter and matter into energy. We have already pointed out that energy is *conserved*. Energy does not transform into anything, and it is only different particles that transform into one another. We shall discuss many examples of such transformation in the following pages. The point is well illustrated by the chemical reaction between carbon and oxygen that we observe in a bonfire. This reaction is

$$C + O_2 \rightarrow CO_2 + \text{photons}.$$

The kinetic energy of photons and CO_2 molecules is produced in this reaction because the combined mass of the C atom and the O_2 molecule is slightly greater than the mass of the CO_2 molecule. This means that while all the energy of the initial ingredients of the reaction is in the form of rest energy, that of the final products is the sum of rest energy and kinetic energy.

Energy is thus conserved but the carriers of it, and the form in which it appears, do change.

Forces and fields

The energy and momentum of a freely moving body do not vary with time. But when two or more bodies interact, their individual momenta (and, in general, their energies) do change. It is not essential that the bodies actually come in contact (collide) for these changes to occur. They can affect each other's motion at a distance.

Thus, for example, the Earth and its satellite attract each other, so that their individual momenta constantly change. The changes in their momenta are equal and opposite in sign, so that the total momentum of the system is conserved. (We perceive the change in the momentum of the satellite, but fail to see the corresponding change in the Earth because the mass of the Earth is enormous compared with that of the satellite, and the change in the velocity of a body for a given change in momentum is inversely proportional to mass.)

The proton and the electron in the hydrogen atom act on each other in nearly the same way. The interaction between the Earth and its satellite is the so-called gravitational, or Newtonian, attraction; and that between the proton and the electron is the electric, or Coulomb, attraction. In both cases, the attractive force is inversely proportional to the square of the separation. The bodies act on each other by producing fields of force around them. Another familiar example of a field of force is the magnetic field, e.g., the geomagnetic field that acts on a compass needle.

A particle placed in a field of force is characterized not only by its rest energy E_0 and kinetic energy T but also by its potential energy U. The total energy is then the sum of three (not two) quantities, i.e. $E = E_0 + T + U$. Potential energy with its sign reversed is equal to the work that must be done to separate two interacting bodies at rest to a distance that is large enough for the interaction between them to be negligible. It follows from this definition that potential energy is negative in the case of attraction.

■ This is the proper place for a digression about the units of energy and mass. The unit of energy in particle physics is the electron volt (eV). It is also convenient to use its derivatives

$$1 \text{ keV} = 10^3 \text{ eV}, \ 1 \text{ MeV} = 10^6 \text{ eV}.$$
$$1 \text{ GeV} = 10^9 \text{ eV}, \ 1 \text{ TeV} = 10^{12} \text{ eV}.$$

The electron volt is the energy gained by an electron traversing a potential difference of one volt. If we recall that

$$1 \text{ joule} = 1 \text{ coulomb} \times 1 \text{ volt}$$

and that one coulomb is the total electric charge of about 6.10^{18} electrons, we readily conclude that

$$1 \text{ eV} \approx 1.6 \cdot 10^{-19} \text{ J}.$$

In elementary-particle physics, the electron volt also serves as the unit of mass. Actually, the unit of mass is the quantity 1 eV/c^2, where c is the velocity of light:

$$1 \text{ eV}/c^2 \approx 1.8 \cdot 10^{-36} \text{ kg}.$$

However, physicists dealing with elementary particles usually regard c as a unit of velocity and prefer to omit c, i.e., to work in a system of units in which $c = 1$. The electron mass is then $m_e \approx 0.511$ MeV, the proton mass is $m_p \approx 938.28$ MeV, and the neutron mass is $m_n \approx$ 939.57 MeV.■

Let us now return to the motion of a body in a central force field that is inversely proportional to the square of the distance from the center of force. According to nonrelativistic mechanics, in the steady motion of a satellite on a circular orbit around the Earth, or of an electron around an atomic nucleus, the absolute value of potential energy is twice the kinetic energy, i.e., $U = -2T$. It is easy to verify this relation. Indeed, the Newtonian potential energy is

$$U = -G_N Mm/r,$$

where r is the distance between the satellite and the center of the Earth (the center of force), m is the mass of the satellite, M is the mass of the Earth, and G_N is Newton's constant ($G_N = 6.7 \cdot 10^{-11}$ $m^3 kg^{-1} s^{-2}$, but the numerical value of G_N is unimportant in this context).

The gravitational force of attraction acting on the above satellite is $F = G_N Mm/r^2$, and the centripetal acceleration is v^2/r. Recalling that the kinetic energy of a satellite is $T = \frac{1}{2}mv^2$, we find that $T = G_N Mm/2r$, so that $T = -\frac{1}{2}U$.

Figure 5 plots U as a function of r and shows the relationship between U and T. The figure also shows the *binding energy* ε, which is defined by

$$\varepsilon = -(U+T).$$

In the case of the Newtonian potential, $T = -\frac{1}{2}U$, so that

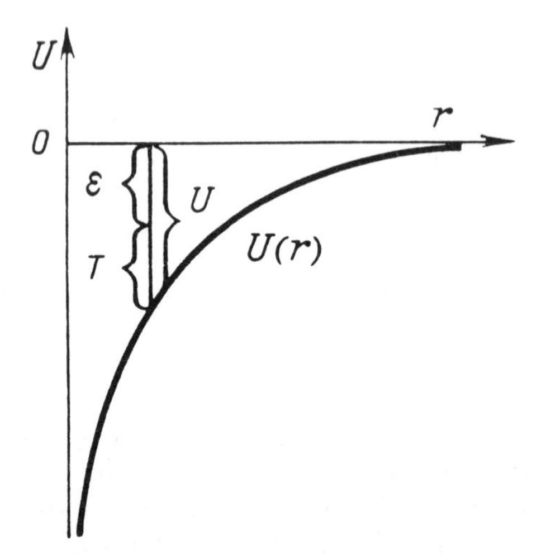

Figure 5. Relationship between the kinetic energy T and potential energy U of a satellite; ε is the binding energy.

$$\varepsilon = -\tfrac{1}{2}U.$$

We see that the mass of the "satellite + Earth" system is less than the sum of the masses of satellite and Earth by the amount ε/c^2. The binding energy of the satellite increases when the satellite is brought closer to the Earth.

Likewise, the mass of the hydrogen atom is less than the sum of the electron and proton masses and also depends on the average separation r between the electron and the nucleus. The corresponding difference between the masses is called the mass defect. When multiplied by c^2, it is equal to the binding energy of the electron.

Quantum phenomena

When describing an atom, we speak of the average separation between an electron and the nucleus and not of the radius of the orbit because, unlike a satellite, the electron is not allowed by quantum mechanics to have a definite orbit.

In contrast to the satellite, the energy of an electron in the atom, and hence the mass of the atom, can assume only discrete (i.e. not continuous) values of energy. This, too, is a requirement of quantum mechanics, which rules the world of the minutest particles of matter.

A physical quantity that plays a very important role in quantum mechanics is the so-called *action* S. The dimension of action is that of the product of energy and time:

$$[S] = [E]\,[t],$$

where brackets [ ] denote the dimension of the quantity they enclose. We know that

$$[E] = [m]\,[l^2]\,[t^{-2}],$$

where l is length and m is mass, and we can readily verify that

$$[S] = [m]\,[l^2]\,[t^{-1}].$$

Just as the velocity of light c is the fundamental constant in relativity, so the quantum of action, $\hbar$ (also called Planck's constant), is the fundamental constant in quantum mechanics:

$$\hbar = 1.0545887(57)\cdot 10^{-34}\ \text{J s}$$

This number clearly shows that in all macroscopic processes the value of S is enormous in comparison with $\hbar$. It is for this reason that macroscopic processes are described so well by classical mechanics and that quantum effects are negligible for them. However, for electrons in atoms, the action S is of the order of $\hbar$, and quantum effects dominate the situation.

One of the striking consequences of quantum mechanics is the quantization of angular momentum. The dimension of angular momentum is readily shown to be that of Planck's constant. In quantum mechanics, the angular momentum of orbital motion can only be an integral multiple of $\hbar$. The discrete nature of angular momentum is not apparent in our daily experience because the angular momenta of macroscopic bodies are truly huge numbers when expressed in units of $\hbar$, and the precision of macroscopic measurements is not sufficient to enable us to detect the discrete character of the angular momentum of, say, a spinning toy top. However, $\hbar$ *is* a natural unit for measuring angular momenta of electrons in atoms. The lowest orbital state of an electron has zero orbital angular momentum, $L = 0$, whilst higher states correspond to $L = \hbar, 2\hbar$, etc.

However paradoxical this may sound, *both* the angular momentum $L = l\hbar$ and its projections onto the coordinate axes are *quantized*. In fact, the projections can assume only integral values (in units of $\hbar$), between $-l$ and $+l$.

An important feature of this quantization is that whilst the projection of the angular momentum of a particle onto one of the coordinate axes, say, the z axis, has a definite value, the projections along the other two axes (x and y) do not. When these projections are measured, any integral value between l and $-l$ can appear with finite probability. We shall turn to the probabilistic nature of quantum mechanics in a moment. But first let us make an acquaintance with a very important notion in particle physics, namely, the notion of spin.

In addition to orbital angular momentum, elementary particles also have an *intrinsic* angular momentum called the spin, which is a multiple of $\frac{1}{2}\hbar$. The spin of the electron and of the proton is $\frac{1}{2}$ (in units of $\hbar$), and the spin of the photon is 1. Particles with half-integral spin (in units of $\hbar$) are called *fermions* and those with integral spin are called *bosons* (after the Italian physicist Fermi and the Indian physicist Bose).

Fermions are "individualistic" while bosons are "collectivistic". Only one fermion with a given spin projection can occupy a given energy level. This is why electrons in atoms do not all occupy the lowest energy level but, as the charge of the nucleus increases, they fill successive atomic shells, thereby forming the Mendeleev Periodic Table. In contrast, bosons tend to occupy the same state.

■ We note, in passing, that this property of bosons is responsible for the superfluidity of helium (the spin of the helium atom is zero). It is also fundamental for lasers.■

The quantization of angular momentum is but one of the numerous manifestations of the quantum nature of sub-microscopic particles.

It must be emphasized that, whilst forcing strict discreteness on a variety of classical quantities (discrete energy levels, quantization of angular momentum), quantum mechanics requires at the same time that a number of other quantities become probabilistic (in contrast to their classical deterministic nature).

An example of this probabilistic character of quantum mechanics was mentioned above in connection with angular momentum. Another example involves the trajectory of a particle. The well-defined trajectory of classical mechanics is replaced in quantum mechanics by the sum over paths. Concepts such as the lifetime of an atomic excited state and the cross section (a quantity with the dimensions of area, characterizing the probability of a process that may result from a collision between particles) have a probabilistic (statistical) interpretation. Particles are described in quantum mechanics by the so-called *wave function*.

In fact, sub-microscopic particles resemble *centaurs*[c] in that they combine the properties of corpuscles (particles) with those of waves. This particle-wave duality is most readily observed for photons. In a photon-electron collision, the photon is as much a particle as the electron: it recoils in a definite direction with a definite momentum, which is balanced by the momentum of the electron. On the other hand, a photon with momentum **p** behaves as if it were a wave with wavelength $\lambda = 2\pi\hbar/|\mathbf{p}|$. The wave properties of photons are particularly well-defined in phenomena such as the diffraction and interference of light.

The same relation between wavelength and momentum, i.e., $\lambda = 2\pi\hbar/|\mathbf{p}|$, is valid not only for photons, but for all other particles: electrons, protons, neutrons, and composite particles such as atoms, molecules, automobiles . . . However, the heavier the body, the greater its momentum, the shorter its wavelength, and, hence, the more difficult it is to detect its wave properties.

The uncertainty relation, which relates the uncertainties in the position and momentum of a particle,

$$\Delta r \, \Delta p \gtrsim \hbar,$$

is a clear expression of a particle-wave duality. The smaller the region in which a particle moves, the greater the uncertainty in its momentum. It is, in fact, this behaviour that results in the existence of the lowest energy state of the electron with nonzero kinetic energy in each atom. This energy state is called the *ground state*. Indeed, for a given size of the atom, the momentum of the electron (and its kinetic energy) cannot be arbitrarily small.

We can use the uncertainty relation to estimate the order of magnitude of the binding energy ε of an electron in the ground state of the hydrogen atom. The potential energy U and the kinetic energy T of the electron are given by

$$U = -e^2/r, \quad T = p^2/2m_e.$$

The uncertainty relation tells us that $p \approx \hbar/r$, and since $2T = -U$ (see Figure 5), we have $\hbar^2/r^2m_e \approx e^2/r$, so that $r \approx \hbar^2/e^2m_e$. This gives the

---

[c]In Greek mythology, a race of half-man and half-horse creatures.

following estimate for binding energy ε:

$$\varepsilon = T \approx e^4 m_e / 2\hbar^2.$$

We are lucky: our crude estimates of r and ε are identical with the exact values of the so-called Bohr radius r_0 and the binding energy ε_0 of the hydrogen atom:

$$r_0 = \hbar^2/e^2 m_e = 0.52917706(44) \cdot 10^{-10}\ \mathrm{m}$$
$$\varepsilon_0 = e^4 m_e/2\hbar = 13.605804(36)\ \mathrm{eV}.$$

In terms of the dimensionless quantity $\alpha = e^2/\hbar c$, we have

$$\varepsilon_0 = \frac{1}{2}\alpha^2 m_e c^2, \quad r_0 = \frac{1}{\alpha}(\hbar/m_e c),$$

where the ratio $\hbar/m_e c = 3.8615905(64) \cdot 10^{-13}$ m is traditionally referred to as the Compton wavelength of the electron. The quantity α is known in atomic physics as the *fine structure constant*. Its value is

$$\alpha = 1/137.03604(11).$$

It is not difficult now to evaluate the velocity of the electron in the hydrogen atom. It turns out to be about 1/137 of the velocity of light, as we have mentioned earlier (see p. 4).

When at atom collides with another atom, or when it absorbs ultraviolet radiation, one of its electrons may be knocked out of the atom (this is called the *ionization* of the atom) or it may be raised to a higher energy level (this is called the *excitation* of the atom). The binding energy in the n-th excited level, ε_n, is related to the binding energy in the ground state, ε_0, by the following formula:

$$\varepsilon_n = \varepsilon_0 (n + 1)^{-2},$$

where n = 1, 2, 3, . . . Of course, discrete energy levels are typical not only for electrons in atoms but also for atoms in molecules (the level spacing is then essentially smaller than in atoms) and for nucleons in atomic nuclei (level spacing much larger than in atoms).

Each molecule, each atom, and each atomic nucleus (except the simplest ones – the proton and the deuteron) thus have a set of discrete excited states in addition to the ground state. It should be clear to the reader that the mass of a molecule, atom, or nucleus in an excited state must be greater than its mass in the ground state.

Atomic and nuclear reactions

We know that, in a fire, the atoms of carbon and hydrogen in the wood react with oxygen atoms in the atmosphere and produce carbon dioxide and water. The sum of the masses of molecules entering the burning reaction is greater than the sum of the masses of resultant molecules. Because of energy conservation, the kinetic energy of the combustion products must be greater than the kinetic energy of molecules entering the reaction. This excess of kinetic energy is the heat released by the burning fuel. It would be wrong to say that the mass is thereby converted into energy. It would be better to say that the mass is partly converted into the kinetic energy of matter and radiation (photons). And it would be perfectly correct to say that energy is partly transformed from one form (rest energy) into another (kinetic energy). We note that the total mass of the system (including matter and photons) remains unaltered.

When solar rays initiate the conversion of carbon dioxide and water in plant leaves into organic compounds and oxygen, the mass of the substance increases, and the energy necessary for this is supplied by the Sun (as the kinetic energy of solar photons).

Ultimately, it is the Sun that has been supplying mankind with usable energy over almost all of its history. But from where does the energy of the Sun itself come? We shall see later that this energy comes from nuclear reactions in which the total mass of nuclei produced in the reactions is lower than the total mass of nuclei entering the reactions. The mass difference (the difference between rest energies) is equal to the excess kinetic energy of the product particles. The Sun emits this energy into the surrounding space, mostly in the form of photons.

Atoms colliding with sufficiently high velocities become excited, i.e., the atomic electrons are raised to excited energy levels, and the mass of the atoms increases. An atom cannot survive for long in an excited state: after a short time, it emits a photon and returns to the ground state. Photons are emitted by atomic electrons falling from one level to another. The important point is that photons emitted by atoms are not stored in advance but are created at the moment of emission. It is the motion of electric charges (electrons) that results in the emission of quanta ("portions") of the electromagnetic field, i.e., photons. Similarly, photons are not stored in the glowing filament of a light bulb: they are created and emitted by "heated-up" electrons.

The energy E of a photon is related to its frequency ω by the formula $E = \hbar\omega$. We recall that the wavelength and frequency of light are related by the formula $\lambda = \omega c$, so that a light quantum with a definite wavelength has a definite energy. The field of an electric charge at rest is purely static and is referred to as the Coulomb field. But the field of a moving charge contains excitations with nonzero frequencies. When the velocity of the charge changes, these excitations are "shaken off", and leave as free photons.

The light emitted by excited atoms is not necessarily visible light. When the atom is heavy, and the excited electrons originate in inner shells, de-excitation results in the emission of x-rays.

Like atoms, excited nuclei emit photons, but the energy of these photons (nuclear γ-ray quanta) is greater than the energy of atomic photons. (We note that the binding energy of the electron in the hydrogen atom is 13.6 eV, whilst the binding energy of a nucleon in a nucleus is, on average, about 8 MeV.)

When their excitation energy is high enough, nuclei can emit particles other than photons. Nuclear reactions of this type are extremely diverse, but can be divided into two large groups. The first comprises reactions in which single nucleons or even clusters of nucleons (nuclear fragments) are ejected from the nucleus. This occurs, for example, in *α-decay* (the α particle is the nucleus of the helium atom and consists of four nucleons) and in the fission of uranium. The second group comprises reactions in which the excess energy of an unstable nucleus is carried off by particles that did not exist in the nucleus prior to the moment of emission. The simplest example of this is the emission of photons. Another example is the emission of a *pair* of particles from the nucleus, namely, an electron and a neutrino. This phenomenon was discovered at the end of the last century and is called *β-decay*.

Weak and strong interaction

It took a long time to identify the nature of the particles emitted in β-decay. One of these particles is electrically charged, but the other is neutral. The charged particle was called the β-particle until it was identified as the electron (the electron itself had been discovered not long before the discovery of β-decay). Soon after the discovery of radioactivity, it was established

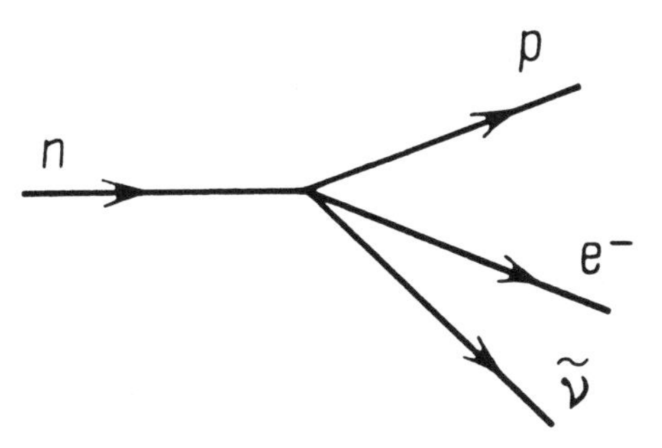

Figure 6. The β-decay of the neutron.

that three types of ray were emitted, namely, α, β, and γ. We now know that α-rays are helium nuclei, β-rays are electrons, and γ-rays are nuclear γ-ray quanta.

In the early 1930s, it became clear that the electron emitted in the β-decay was not alone but was accompanied by another particle that was electrically neutral. This particle was given the name neutrino (Italian for a "small neutron").

The simplest example of β-decay is the decay of the free neutron (Figure 6), in which the neutron transforms into a proton, an electron, and a neutrino (in fact, an antineutrino[c]; the meaning of the prefix "anti" will be explained a little later):

$$n \rightarrow p + e^- + \tilde{\nu}.$$

The decay of the neutron is possible because its mass exceeds the sum of the masses of the proton, electron, and antineutrino. As in the emission of a γ-ray photon by an excited nucleus, the particles emerging from the β-decaying neutron are not present in it but are created at the moment of decay and are "shaken off". But whilst only one particle, the photon, is emitted when the state of an atomic electron changes, the neutron-to-proton transition results in the simultaneous emission of *two* particles, i.e., an electron and an antineutrino.

From the standpoint of energy and momentum conservation, β-decay is no different from the other processes discussed earlier in this book.

---

[c]The neutrino is usually represented by the Greek letter ν (nu). The antineutrino is indicated by a tilde or bar: $\tilde{\nu}$ or $\bar{\nu}$.

However, it does introduce us to a new type of fundamental force that has not been encountered in our story before. We have already mentioned the *gravitational* interaction. We have also discussed several manifestations of the *electromagnetic* interaction, e.g., the attraction of charged particles of unlike sign, and the emission and absorption of photons. We have also touched upon, albeit implicitly, the so-called *strong* interaction, which is responsible for the attraction between nucleons in the nucleus. This interaction is called "strong" because nuclear forces are much stronger than electromagnetic forces (this follows from the higher binding energy of nucleons in the nucleus).

In β-decay, we encounter a manifestation of the fourth type of fundamental force, the so-called *weak* interaction. It is said to be "weak" because, first, its effects seem to be utterly negligible in our daily experience, and, second, the interaction is much weaker in atoms and nuclei than the strong and electromagnetic interactions. The processes due to it have lower probability and thus proceed at a slower rate.

> This is a good place for a brief digression. It is well-known that the magnetic field does not deflect γ-rays, but α- and β-rays are deflected by it in opposite directions. This is illustrated by the symbolic α-β-γ drawing on p. vi. I recall one of the lengthy evening discussions about the future of physics that the head of the Theoretical Department of the Institute of Theoretical and Experimental Physics, Professor I.Ya. Pomeranchuk, used to have with his students and colleagues. Professor V.B. Berestetskiĭ, famous for this work on quantum electrodynamics, pointed out during the discussion that the above drawing, commonly found in school physics textbooks, symbolizes the three fundamental interactions: α-decay is a manifestation of the strong, β-decay of the weak, and γ-decay of the electromagnetic interactions. During the first half of this century, the physics of each of these three interactions grew into a separate science. However, we shall see later that a synthesis of these sciences is now taking place.

Let us now return to our discussion of β-decay. It may appear at first glance that the world in general and mankind in particular could survive without the weak interaction. Indeed, β-decay would seem to be a relatively exotic phenomenon. However, this conclusion would be profoundly wrong. Suffice it to say that if it were possible to "turn off" the weak interactions, our Sun would be extinguished.

The point is that the key process that opens the way to all further nuclear reactions in the Sun is the transformation of two protons and one electron

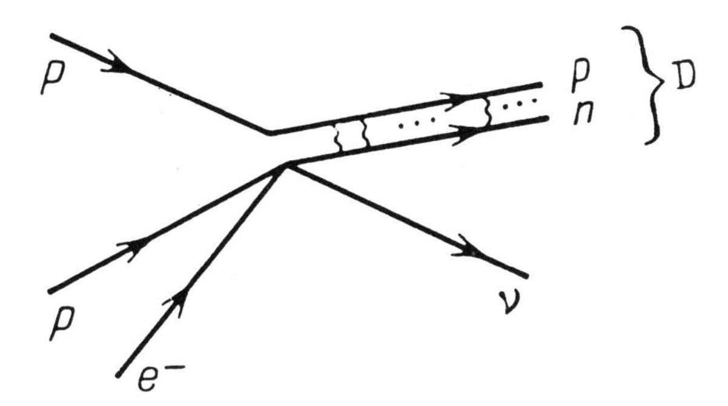

Figure 7. Weak reaction $p + p + e^- \rightarrow D + \nu$.

into a deuteron D and a neutrino ν.

Note that the single-step transformation (Figure 7)

$$p + p + e^- \rightarrow D + \nu$$

takes place in only 0.25% of all cases. In all other cases, the reaction proceeds in two stages. The first step is the creation of a positron e^+ in the reaction (Figure 8).

$$p + p \rightarrow D + \nu + e^+.$$

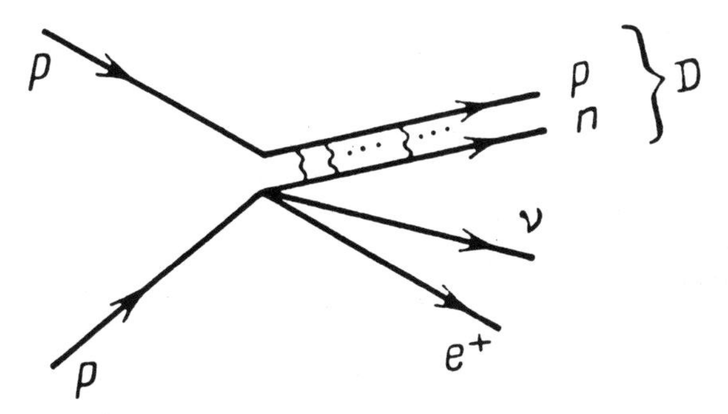

Figure 8. Weak reaction $p + p \rightarrow D + \nu + e^+$.

The second step involves the annihilation of the positron and one solar electron:

$$e^+ + e^- \rightarrow 2\gamma \text{ or } 3\gamma.$$

Positrons and annihilation will be discussed in more detail later (see the section entitled *Antiparticles*).

We recall that the deuteron D, i.e., the bound state of a proton and a neutron, is the nucleus of deuterium. The wave lines in Figures 7 and 8 are meant to symbolize the strong nuclear interactions tying the proton and the neutron into a deuteron. Now, the deuteron binding energy is roughly 2.2 MeV, the difference between neutron and proton masses is 1.3 MeV, the positron mass is 0.5 MeV, and the neutrino mass is negligible (or may even be zero), so that we can readily find the energy released in the process shown in Figure 9: it is a mere 0.4 MeV.

The weak process described above, which in a sense can be regarded as the inverse of the β-decay of the neutron, is the main source of solar neutrinos.

However, we have just seen that the kinetic energy released in this process is relatively low. Most of the release of heat is due to the *fusion* of two deuterium nuclei into a helium nucleus consisting of two protons and two neutrons. This proceeds mainly through the following two reactions:

$$D + p \rightarrow {}^3He + \gamma + 5.5 \text{ MeV},$$
$${}^3He + {}^3He \rightarrow {}^4He + 2p + 12.9 \text{ MeV}.$$

The first involves both the strong and the electromagnetic interaction (a γ-ray is emitted), and the second involves the strong interaction alone. The considerable energy release in the second reaction occurs because the nucleons in an α particle are closely packed and have a high binding energy.

Such fusion reactions are also called *thermonuclear reactions* because they can be sustained only at very high temperatures. Such temperatures are required in order to ensure that the nuclei approach one another as closely as possible. Since electric charges of the same sign are known to repel, enough energy must be supplied to the nuclei to overcome this repulsion and allow them to reach a separation of the order of 10^{-15} m.

The fusion reaction is at once the great hope *and* the great danger for mankind, threatening its very survival. If it were possible to exploit controlled thermonuclear fusion for industrial purposes, vast energy resources would become available, and the threat of an energy crisis would be eliminated. On the other hand, if the huge stock of hydrogen bombs that has now accumulated (and keeps on growing at progressively higher rates in the nuclear arsenals of an increasing number of countries) were to explode, mankind would be wiped out.

High-energy physics

So far, we have dealt only with preliminaries. The subject we are to study, i.e., high-energy physics, has nothing to do with nuclear power stations or hydrogen bombs.

High-energy physics explores the nature of fundamental forces and the structure of elementary particles. The lengthy introduction devoted to atomic and low-energy nuclear physics was necessary for two reasons. First, an advanced study should be preceded by an elementary one. Second, high-energy processes have much in common with processes occurring at low energies. When compared with what we now regard as "high energy", the energy of the nuclear fusion reaction is as low as the energy of visible light in comparison with that of nuclear γ-rays). Thus, no matter how many particles are born in a high-energy reaction, the reaction will obey the law of conservation of energy. A new heavy particle can therefore be created only in a collision between sufficiently energetic projectiles. This is the reason we have spent so much time on the preceding pages, discussing processes in which lighter particles transform into heavier species and vice versa. High-energy processes are no exception in this respect. But the conservation of energy and momentum is, so to say, the routine side of high-energy physics. Its real fascination lies in that it has opened to us a world of fundamental principles and phenomena that are at the same time profound and surprising.

The first stage in the history of high-energy physics, between the early 1930s and the late 1940s, was mostly concerned with *cosmic rays*. Primary cosmic rays consist of fast protons arriving on the Earth from outer space. They collide with the nuclei of atoms in the atmosphere and thereby create numerous secondary particles. It was found that these secondary particles included not only the familiar photons, electrons, and nucleons, but also some totally unfamiliar species. Several generations of ever-larger accelerators of charged particles have been constructed since the beginning of the 1950s in order to determine the nature of these new species of particles.

Accelerators

Depending on the type of accelerated particle, accelerators are classified as proton or electron accelerators. An accelerator can be of circular or linear design. At present, there are many more working circular accelerators than linear machines.

One of the largest circular proton accelerators was built by the European Organization for Nuclear Research (CERN) near Geneva in Switzerland. Another major machine is at the Fermi National Laboratory in Batavia, not far from Chicago in the U.S.A. The maximum energy of protons accelerated in these two machines is 400 GeV and 1000 GeV, respectively. The particles in Geneva and Batavia are accelerated in circular tunnels with a circumference of about seven kilometers. Before the big accelerator near Geneva was commissioned, the highest recorded energy (76 GeV) was achieved by the proton accelerator of the Institute of High Energy Physics at Protvino, near Serpukhov, in the USSR. This machine became operational in 1967. The circumference of the circular beam tunnel of this accelerator is about 1.5 km.

Electromagnets installed along the entire tunnel of the circular accelerator force the particles into circular paths in the evacuated beam tube. This doughnut-shaped tube is called the vacuum chamber. The stronger the magnetic field in the magnets, the more energetic the particles that can be kept revolving in the chamber. Superconducting magnets were installed in the Batavia accelerator in 1983, raising the maximum energy of accelerated protons to 1000 GeV.

Magnets thus confine the particles to a circular "race-track" and an electric field plays the role of the accelerating "whip". Several accelerating gaps in which the electric field is produced are arranged along the ring. A particle completes a large number of revolutions before it acquires the required energy, and the electric field in the gaps of a circular accelerator need not be very strong. On the other hand, the accelerating electric field in a linear accelerator must be as high as possible because the particles have to gain all their energy in a single passage through the system.

Record values of the alternating electric field have been achieved at the Institute of Nuclear Physics at Novosibirsk. A field of the order of 10 million volts was achieved in a 10 cm section of a future linear electron accelerator in which the rate of acceleration will be about 100 MeV/m.

The feasibility of using lasers to achieve still higher acceleration rates is being actively discussed. However, this technology belongs to the next century.

The largest working linear accelerator is at Stanford, near San Francisco. It is just over 3 km long and accelerates electrons to 20 GeV. The two largest circular electron accelerators produce maximum energies almost as high. One of them is also at Stanford, and the other is near Hamburg, in West Germany. The circumference of these accelerators is just over 2 km.

The attentive reader may have noticed that the acceleration efficiency per unit length is higher in proton accelerators than in circular electron machines. This is so because electrons, being lighter particles, emit more intensively the so-called synchrotron radiation as they move on a curved trajectory. Energy lost through the emission of synchrotron radiation is reduced by reducing the centripetal acceleration, i.e. by increasing the radius of the electron accelerator.

When the particles have reached the required energy, they are extracted in the form of a beam and sent on to a target in which they collide with the target nuclei and create new particles. Some of these new particles have long lifetimes and escape from the target, whilst others are so short-lived that they decay inside the target (some of them perish even before they leave the atom in whose nucleus they were born). In this latter case, the particles emerging from the target are the decay products. They are focussed by special magnets into secondary beams and are directed into experimental areas containing equipment that detects them as well as their collisions and decays.

The accelerators that have become increasingly important in recent years are the circular machines in which an accelerated beam of particles is made to collide not with a fixed target but with a beam of particles accelerated in the opposite direction. The particular advantage of this colliding-beam technique is the considerable gain in the fraction of energy that can be used for the creation of new particles.

Consider two colliding beams of particles of mass m, and energy E, and opposite momenta ($+\mathbf{p}$ and $-\mathbf{p}$). The total energy of the two colliding particles is 2E, and the total momentum is zero. The coordinate system in which the total momentum of the two particles vanishes is called the center-of-mass reference frame. In this particular case, the center-of-mass frame is the laboratory reference frame. The energy 2E corresponds to a mass $M = 2E/c^2$ and, in principle, the whole of this mass can go into the creation of new particles.

Now consider the collision of the same beam with a *fixed* hydrogen target (a target containing hydrogen atoms). Let the energy of each particle in the beam be E, as before, the mass and momentum being m and **p**, so that $\mathbf{p}^2c^2 = E^2 - m^2c^4$. We shall use μ to denote the proton mass (in the hydrogen target). By definition, the mass of the "particle + proton" system, i.e., the total energy of the particle and the proton in the center-of-mass frame, is given by

$$M^2c^4 = (E + \mu c^2)^2 - \mathbf{p}^2c^2 = 2E\mu c^2 + \mu^2c^4 + m^2c^4.$$

In this case, the center-of-mass frame moves relative to the laboratory frame. If E is much greater than μc^2 and mc^2, the energy of the colliding particles in the center-of-mass frame in the first case is greater by the factor of $\sqrt{(2E/\mu c^2)}$ than in the second. Actually, it is only the center-of-mass energy that constitutes the effective energy of collision and determines the collision's characteristics.

No formulas are needed to realize that the head-on collision of two automobiles is much more energetic than the collision of one of them with a parked car. However, the gain in energy is much higher for relativistic particles than for nonrelativistic cars.

The first colliding-beam accelerators (called *colliders*) were built in the 1950s, but they have produced the most interesting results only during the last decade. Some of the collider experiments will be discussed later, but first let us try to summarize the main outcome of high-energy experiments. The most spectacular achievements of high-energy physics are the antiparticles, the hadrons and quarks, the generations of leptons and quarks, broken symmetries, and the fundamental vector bosons. In the rest of the book I will explain what these words mean.

Antiparticles

The first antiparticle, the *positron*, was predicted theoretically and then discovered experimentally in the early 1930s. The positron is the antiparticle of the electron. Its mass and the absolute value of its charge are the same as those of the electron, but the positron has a positive charge (in contrast to the negative charge of the electron). The electron and the positron are therefore represented by e^- and e^+ respectively.

In empty space, the positron is as stable as the electron. However, the electron-positron encounter ends badly for both: they "disappear", i.e., annihilate each other, emitting photons (γ-rays). As a rule, two or three γ-ray photons are emitted in each electron-positron annihilation:

$$e^+ + e^- \to \gamma + \gamma \quad \text{or} \quad e^+ + e^- \to \gamma + \gamma + \gamma.$$

There is nothing mystical about the "disappearance" of an electron and a positron. All that happens is that, in contrast to the reactions discussed earlier, the rest energy of the electron and positron is completely transformed in this annihilation into the kinetic energy of the γ-ray photons.

The inverse of the electron-positron annihilation process is also observed in laboratory (accelerator) experiments. When two hard γ-ray photons collide, an electron-positron *pair* is created:

$$\gamma + \gamma \to e^+ + e^-.$$

The discovery of the positron was followed by the discoveries of other antiparticles. The *antiproton* and *antineutron*, and even light *antinuclei*, have been produced in accelerators since the middle 1950s. Antiparticles are usually represented by the same letters as the corresponding particles, but with a tilde (or bar) placed above them. For example, $\tilde{p}$ is the antiproton, $\tilde{n}$ is the antineutron, and $\tilde{\nu}$ is the antineutrino.

The mass of each antiparticle is strictly equal to the mass of the corresponding particle, but their charges have opposite signs. The imagined particle → antiparticle replacement operation is called *charge conjugation*. This operation transforms the photon, which has no electric or other charge, into itself (examples of "other charges" will be given in a moment). The photon belongs to a comparatively rare species: it is truly neutral and has no "antiparticle" of its own.

It is natural to ask why it is that electrons and positrons can mutually annihilate into photons, whilst electrons and protons cannot. Why is the hydrogen atom stable? What is it that forbids the reaction $e^- + p \to 2\gamma$? Obviously, if this reaction were allowed, the world would decay into photons and neutrinos (neutrinos from neutron decays). Surely, a disagreeable prospect!

The stability of hydrogen is a hint that the electric charge cannot be the only conserved charge (the only conserved quantum number). The conservation of baryonic and leptonic charges has been proposed as a way of explaining the stability of hydrogen and of heavier atoms, as well as the absence of a number of other processes.

Let us begin with the baryonic charge. There is a large family of particles called *baryons* (from the Greek *barys*, meaning heavy). According to the baryonic charge hypothesis, each baryon is assigned a unit baryonic charge. The proton is the lightest baryon. Scores of other, heavier, baryons are also known. Each baryon has its own antiparticle, namely, the corresponding *antibaryon* with the negative unit baryonic charge. (It then follows that, although the neutron is electrically neutral, it is not a truly neutral particle.)

The family of particles called *leptons* (from the Greek *leptos*, meaning slender, small; *lepta* is a small coin) is considerably smaller than the baryon family. The lightest among charged leptons, the electron, is assigned a leptonic charge of $+1$. According to the leptonic charge hypothesis, the neutrino has the same leptonic charge as the electron. The positron and the antineutrino each have leptonic charge of -1. It is readily verified that both baryonic and leptonic charges are conserved in the neutron decay

$$n \rightarrow p + e^- + \tilde{\nu}$$

The precision of baryonic and leptonic charge conservation laws will be discussed in the final chapters.

Let us now address the question of whether the world surrounding us contains antiparticles.

Particles and antiparticles cannot coexist in close proximity because of annihilation reactions. The few antiparticles that are produced in laboratory experiments perish sooner or later as they come into contact with their hostile environment. However, antiworlds built of antiatoms could well exist in remote regions of the Universe, far from our ordinary matter. The energy levels of antiatoms and atoms are identical, and their chemical properties are indistinguishable (very slight differences between matter and antimatter are detectable only in weak interactions). It follows that, in principle, there could be "antilife", "antipeople", and "antiworlds". The photons that would come to us from antistars would be indistinguishable from photons arriving from ordinary stars. Optical observations could not therefore distinguish between a star and an antistar. In principle, advances in neutrino astronomy may help in this respect. Indeed, ordinary stars, such as our Sun, emit neutrinos from nuclear fusion reactions; antistars would emit antineutrinos.

Astrophysicists are quite skeptical about the possible existence of antiworlds. One of the reasons for this is that antiprotons have not been detected in appreciable numbers in primary cosmic rays from remote parts of the Universe. Another argument is that the characteristic γ-ray photons with energy equal to the electron rest energy, which should be emitted as a result of the annihilation of slow electrons with positrons at the matter-antimatter interface ($e^+e^- \rightarrow 2\gamma$), have not been observed.

The problem of why our world is composed of matter and not of antimatter, or why it is not a mixture of equal amounts of both, is increasingly attracting the attention of theorists. Meanwhile, experimentalists have put positron and antiproton beams to work in their experiments. In fact, most existing colliders send a beam of particles against a beam of the corresponding antiparticles: protons against antiprotons, electrons against positrons.

Hadrons and quarks

The study of strong interactions has exposed a whole new stratum of experimental and theoretical physics. It became clear as early as the 1940s that nucleons are definitely not the only strongly-interacting particles. They belong to an extensive class of particles that were later called *hadrons* (from the Greek *hadros*, meaning massive, robust, heavily built). A cornucopia of new hadrons was created with the advent of big accelerators: more than three hundred hadrons are now known.

It was suggested in the middle 1960s that all hadrons consist of more fundamental particles, christened *quarks*. Subsequent studies confirmed the validity of this hypothesis.

Each quark has a spin of $\frac{1}{2}$. The existence of five species of quark has been confirmed so far. They are designated u, d, s, c, and b (they are listed here in the order of increasing mass: $m_u \approx 5$ MeV, $m_d \approx 7$ MeV, $m_s \approx 150$ MeV, $m_c \approx 1.3$ GeV, $m_b \approx 5$ Gev). The existence of the sixth, heavier quark, t, is anticipated. The failure to find hadrons incorporating a t quark indicates that $m_t > 20$ GeV.[⊆]

The u, c, and t quarks each have an electric charge of +2/3, while the

[⊆]Several events involving the production and decay of t quarks were reported in the summer of 1984, indicating that $m_t \approx 40$ GeV. However, we cannot be sure as yet that the t quark has really been discovered.

d, s, and b quarks have a charge of $-1/3$; the two groups are called the upper and lower quarks, respectively. The letters denoting quarks come from the words *up, down, strange, charm, bottom*, and *top*.

The quark model was proposed when only the so-called light hadrons, i.e., hadrons consisting of only the light u, d, and s quarks, were known and it immediately brought order to the whole systematics of these hadrons. Not only has it proved possible to understand the structure of the particles known at that time, but a number of further hadrons were correctly predicted.

All hadrons can be divided into two large classes. Those in the first class are called the baryons and consist of three quarks. Baryons (already mentioned in the preceding section) are *fermions*: they have half-integral spins. Those in the second class are called *mesons* and consist of a quark and an antiquark. Mesons are *bosons*: they have integral spins (bosons and fermions were discussed earlier, under the heading *Quantum phenomena*).

We have already pointed out that nucleons are the lightest baryons. The proton consists of two u quarks and one d quark (p = uud). The neutron consists of two d quarks and one u quark (n = ddu) and is heavier than the proton because the d quark is heavier than the u quark. But by comparing the masses of nucleons and quarks, you immediately notice that the nucleon masses exceed by almost two orders of magnitude the sum of the masses of the three corresponding quarks. This occurs because nucleons consist not of "bare" quarks but of quarks wrapped up in a sort of heavy gluon "coat" (*gluons* are introduced in the next Section).

Baryons containing quarks other than u and d are called *hyperons*. For example, the lightest of hyperons, the Λ-hyperon, consists of three species of quarks: Λ = uds.

The lightest mesons are the π-mesons, or *pions*: π^+, π^-, π^0. The quark structure of charged pions is simple:

$$\pi^+ = u\tilde{d},\ \pi^- = d\tilde{u}.$$

The neutral pion is a linear combination of the states $u\tilde{u}$ and $d\tilde{d}$: π^0 spends half its time in the state $u\tilde{u}$ and the other half in the state $d\tilde{d}$. The π^0 meson can be found with equal probability in each of these states. The mass of the π^+ and π^- mesons (these mesons are each other's antiparticles) is approximately 140 MeV. The mass of the π^0 meson (like the photon, the π^0 meson is truly neutral) is approximately 135 MeV.

In the order of increasing mass, pions are followed by the K mesons, whose mass is about 500 MeV:

$K^+ = u\tilde{s}$, $K^0 = d\tilde{s}$, $\tilde{K}^0 = s\tilde{d}$, $K^- = s\tilde{u}$.

The K^+ and K^- mesons are each other's antiparticles. The same is true for the K^0 and $\tilde{K}^0$ mesons, so that they are not truly neutral particles.

Note that particles containing s quarks are said to be *strange*, and the s quark itself is called the *strange quark*. This term arose in the 1950s when some properties of the strange particles puzzled the physicists.

Obviously, nine different states can be constructed out of the three quarks (u, d, s) and three antiquarks ($\tilde{u}$, $\tilde{d}$, $\tilde{s}$):

$$\begin{array}{ccc} u\tilde{u} & u\tilde{d} & u\tilde{s} \\ d\tilde{u} & d\tilde{d} & d\tilde{s} \\ s\tilde{u} & s\tilde{d} & s\tilde{s} \end{array}$$

Seven out of these nine states (three pions and four kaons) were discussed above; the remaining two are the superpositions, i.e. linear combinations[⊆], of the states $u\tilde{u}$, $d\tilde{d}$, and $s\tilde{s}$. The mass of one of the two is 550 MeV (this is the η meson) and that of the other is 960 MeV (this is the η' meson). Like the π^0 meson, the η and η' mesons are truly neutral particles.

The nine mesons we have just discussed have zero spins: J = 0. Each such meson consists of a quark and an antiquark with zero orbital angular momentum, L = 0. The spins of the quark and the antiquark point in opposite directions, so that their resultant spin is also zero, S = 0. The spin J of a meson is the algebraic sum of the orbital angular momentum **L** and the total spin **S** of the quarks:

$$\mathbf{J} = \mathbf{L} + \mathbf{S}.$$

The sum of two zeros in our case naturally yields zero.

Each of the above nine mesons is the lightest of its ilk. For example, consider the mesons in which the orbital angular momentum of the quark-antiquark pair is **L** = 0, as before, but the quark and antiquark spins are parallel, so that **S** = 1 and, hence **J** = 1. Such mesons form the heavier nonet (group of nine)

[⊆]Further details concerning quantum-mechanical superpositions are given on p. 42.

$\varrho^+, \varrho^-, \varrho^0$	$K^{*+}, K^{*0}, K^{*-}, \tilde{K}^{*0}$
770 MeV	892 MeV
ω^0	ϕ^0
783 MeV	1020 MeV

There are numerous mesons for which L ≠ 0 and J ·ᒪ 1 (a meson with the record spin J = 6 was discovered in 1983 on the Serpukhov accelerator).

We now turn to baryons consisting of u, d, and s quarks. According to the quark model, the orbital angular momenta of the three quarks in a nucleon are zero, and the spin **J** of the nucleon equals the algebraic sum of the quark spins. The spins of the two u quarks in the proton are parallel, whilst that of the d quark points in the opposite direction. The result is that the proton has $J = \frac{1}{2}$.

In the quark model, the proton, the neutron, the Λ hyperon, and five further hyperons form an octet (a group of eight) of baryons with $J = \frac{1}{2}$, while baryons with $J = \frac{3}{2}$ form a decuplet (a group of ten):

Quarks		Particles	Mass
ddd udd uud uuu		Δ^- Δ^0 Δ^+ Δ^{++}	1232 МэВ
dds uds uus	⟵⟶	Σ^- Σ^0 Σ^+	1385 МэВ
dss uss		Ξ^- Ξ^0	1530 МэВ
sss		Ω^-	1672 МэВ

The Ω hyperon – the vertex of this inverted pyramid – was found experimentally in 1964. Its mass proved to be just that predicted by the quark model.

However, the real triumph of the quark model came with the discovery of *charmed* particles containing c quarks (c for charm). The first charmed particle, the so-called J/ψ meson with a mass of 3.1 GeV, was discovered in 1974 (this particle is sometimes said to have *hidden charm* because it consists of the pair $c\tilde{c}$). The J/ψ meson was discovered almost simultaneously on two accelerators. A group of experimentalists working on an electron-positron collider found the J/ψ in the reaction $e^{\pm}e^- \rightarrow J/\psi$. Another group, working with a proton accelerator, found the J/ψ among the products of the collision between the proton beam and a beryllium target. It was detected by observing $J/\psi \rightarrow e^+e^-$ decays. The first group named the meson ψ, and the second named it J. Hence the double symbol J/ψ.

The J/ψ meson is one of the states of the $c\tilde{c}$ system, called *charmonium*. The $c\tilde{c}$ system is in some ways similar to the hydrogen atom. However, whatever the state of the latter (whatever the level to which its electron has been raised), we always refer to it as the hydrogen atom. In contrast, the different levels of charmonium (and of other quark systems, too) are treated as distinct mesons. About ten such mesons (charmonium levels) have now been discovered and analysed. These levels differ by the relative orientation of the quark and antiquark spins, the magnitude of orbital angular momentum, and the radial properties of wave functions.

The discovery of charmonium was followed by the discovery of mesons with explicit "naked" charm:

$D^+ = c\tilde{d}$	$D^0 = c\tilde{u}$	$F^+ = c\tilde{s}$
$D^- = d\tilde{c}$	$\tilde{D}^0 = u\tilde{c}$	$F^- = s\tilde{c}$
1869 MeV	1865 MeV	2020 MeV

(rounded-off figures are given here for the masses of charmed mesons). Charmed baryons have also been discovered.

The discovery and theoretical interpretation of charmed particles and, later, of still heavier hadrons containing b quarks, provided a spectacular confirmation of the quark theory of hadrons. Because the c and b quarks had such large masses, the picture of the energy levels of the quark-antiquark system was revealed in all its richness. The psychological impact of this discovery was immense. Even those who had been extremely skeptical about quarks had to accept their reality.

Confinement of quarks

As all hadrons consist of quarks, it seems logical to expect the existence of free quarks. Free quarks should be easy to detect because of their fractional electric charges. A fractional charge cannot be neutralized by any number of electrons or protons: an "undershoot" or "overshoot" is inevitable. If, say, an oil droplet contains a single quark, the charge of the droplet will be fractional. Experiments with droplets were conducted as early as the beginning of this century to determine the electron charge. They were recently repeated, as part of a search for quarks, with much higher precision. No fractional charges were found. A very accurate mass-spectrometer analysis of water gave $\sim 10^{-27}$ as the upper limit for the ratio of the number of free quarks to the number of protons, i.e. the same negative result.

■ True, experimentalists at a Stanford University laboratory detected fractional charges in small niobium spheres suspended in magnetic and electric fields, but these results have not been confirmed in other laboratories.■

Most specialists are now inclined to conclude that nature does not allow quarks to exist in the free state.

A paradoxical situation has thus arisen. Quarks undoubtedly exist inside hadrons. This is evidenced not only by the above quark systematics of hadrons but also by the direct probing of the interior of nucleons with high-energy electrons. A theoretical analysis of this process (called *deep-inelastic scattering*) shows that electrons are scattered inside hadrons by point-like particles with charges equal to $+\frac{2}{3}$ and $-\frac{1}{3}$, and spin $\frac{1}{2}$. An electron undergoing deep-inelastic scattering sharply changes its momentum and energy, transferring much of them to a quark (Figure 9). In principle, this is very reminiscent of the behavior of α-particles impinging on an atomic nucleus (Figure 10). It was in just this way that the existence of atomic nuclei was established in Rutherford's laboratory at the beginning of the century.

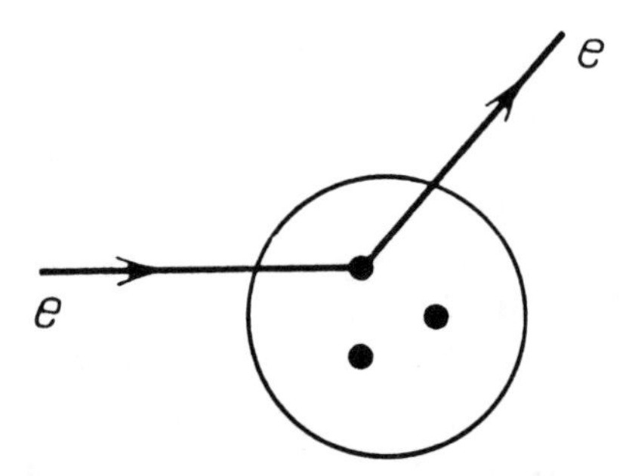

Figure 9. Scattering of an electron by one of the three quarks in the proton. The large circle represents the proton, the black dots represent the quarks.

The fractional charges of quarks manifest themselves in a further deep-inelastic process, namely, in the production of hadron jets in e^+e^- annihilation at high energy (in large colliders). We shall return to hadron jets in e^+e^- annihilation at the end of the book.

Quarks are thus seen definitely to be present within hadrons, but it is impossible to set them loose. This is called *quark confinement*. A quark

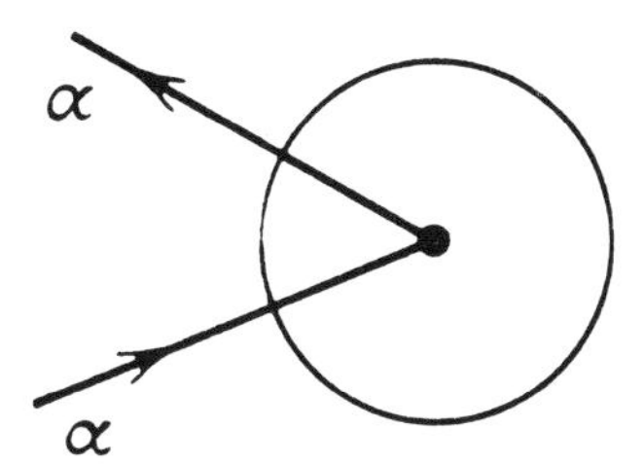

Figure 10. Scattering of an α particle by an atomic nucleus. The large circle represents the atom, and the black dot at the center represents the nucleus.

that has received energy in a collision with an electron (see Figure 9) does not fly out of the nucleon as a free particle but instead dissipates its energy by forming quark-antiquark pairs, i.e., new hadrons (mostly mesons).

Any attempt to break a meson into its constituent quark and antiquark is like an attempt to break a compass needle into north and south poles: by breaking the needle, we get two magnetic dipoles. By breaking a meson, we obtain two mesons. The energy expended in pulling apart the initial quark and antiquark goes into creating a new antiquark + quark pair, with the two pairs forming two mesons.

Actually, the analogy with the magnetic needle is both incomplete and deceptive. Indeed, we know that iron contains no magnetic poles, either at the microscopic or macroscopic level; magnetic dipole moments are entirely due to the spin and orbital motion of electrons. In contrast, individual quarks do exist deep inside hadrons, and we observe them better as we penetrate deeper into a hadron.

The familiar picture in gravitation and electrodynamics is that the forces between particles grow as the particles are brought closer, and diminish when they are pulled apart (1/r-type potentials). The situation with the quark and antiquark is different. There is a critical radius $r_0 \simeq 10^{-15}$m, such that, for $r \ll r_0$, the potential between a quark and antiquark is more or less similar to the Coulomb or Newtonian potential, but its behavior changes dramatically for $r \gtrsim r_0$: it *increases*.

There are reasons for expecting that, if the world contained the heavy quarks (c, b, t) but not the light quarks (u, d, s), the potential would increase linearly with r from $r \simeq r_0$ onward, so that confinement could be described by a funnel-type curve (Figure 11; cf. Figure 5). A linearly increasing potential corresponds to a force independent of distance. We

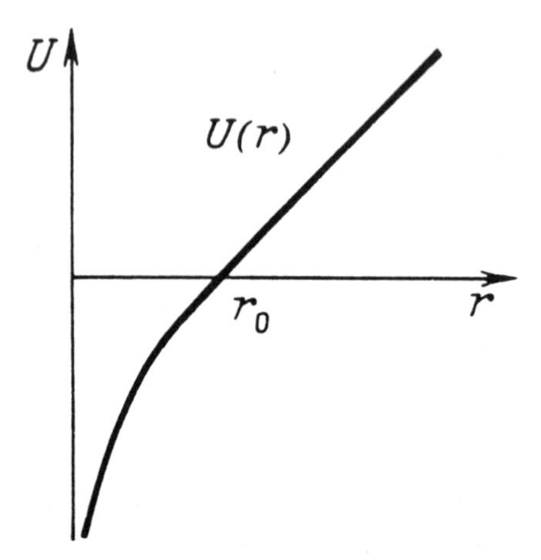

Figure 11. Funnel-like potential describing the confinement of quarks in hadrons.

recall that the potential energy of an ordinary stretched spring increases quadratically with elongation. This is why confinement by a linearly increasing potential can be called *soft*.

Unfortunately, the production of pairs of light quarks does not enable a given quark and antiquark to separate in the real world to more than 10^{-15} m without being confined again, this time in two different mesons. This means that we cannot test the soft spring of confinement at larger distances.

But what are the forces that give quarks this truly unique behavior? What is the extraordinary glue that makes them stick together?

Gluons. Color

The strong field of force produced by quarks and antiquarks, and acting on them, is called the *gluon field*. The particles that constitute the excitation quanta of this field are called *gluons* and are usually denoted by the letter g. Gluons are to the gluon field what photons are to the electromagnetic field. Gluons have the same spin as photons, i.e., $J = 1$ (as always, in units of $\hbar$). Gluons, as photons, have negative parity, $P = -1$ (parity will be introduced in the section entitled *C, P, T symmetries*).

Particles with unit spin and negative parity ($J^P = 1^-$) are called vector particles because their wave functions transform under rotations and reflections of coordinates as ordinary space vectors. The gluon, like the photon, can therefore be classified as a fundamental *vector boson*.

The theory of interactions between photons and electrons is called quantum electrodynamics. The theory of interactions between gluons and quarks is called quantum chromodynamics (from the Greek, *chromos*, meaning color). This is the first time the term "color" appears on the pages of this book. I shall try to explain what stands behind it.

We already know that five distinct quark species (the standard phrase is *quark flavors*) u, d, s, c, b have been observed, and the discovery of the sixth flavor (t) is anticipated. However, quantum chromodynamics holds that each of these quarks is not one but three distinct particles. The number of quarks is thus not six but eighteen or, rather, thirty-six if antiquarks are taken into account. It is usually said that the quark of each flavor exists in three incarnations of different color. As a rule, the quark colors are: yellow (y), blue (b), and red (r). The colors of antiquarks are antiyellow ($\tilde{y}$), antiblue ($\tilde{b}$), and antired ($\tilde{r}$). These names are, of course, mere conventions and have nothing to do with the familiar optical colors. They are only a convenient notation for the specific quark charges. These charges are the sources of the gluon fields in the same way that the electric charge is the source of the photon field.

It was no slip of the tongue (or, rather, pen) when I used plural for the gluon fields and singular for the photon field. The point is that there are eight color species of gluons. Each gluon carries two charges: a color charge (y, b, or r) and an "anticolor" charge ($\tilde{y}$, $\tilde{b}$, or $\tilde{r}$). Altogether nine pair combinations can be formed from the three colors and three "anticolors":

$$\begin{matrix} y\tilde{y} & y\tilde{b} & y\tilde{r} \\ b\tilde{y} & b\tilde{b} & b\tilde{r} \\ r\tilde{y} & r\tilde{b} & r\tilde{r} \end{matrix}$$

These nine pair combinations naturally divide into six off-diagonal *explicitly colored* pairs, namely,

$y\tilde{b}$, $b\tilde{y}$, $b\tilde{r}$, $r\tilde{b}$, $r\tilde{y}$, $y\tilde{r}$

and three diagonal pairs (found on the diagonal of our table) with a sort of *hidden color*:

$y\tilde{y}$, $b\tilde{b}$, $r\tilde{r}$.

Like the electric charge, color charges are conserved. This is why the six off-diagonal explicitly colored pairs cannot mix with one another. As for the three diagonal pairs with hidden color, the conservation of color charges does not forbid the transitions

$$y\tilde{y} \leftrightarrows b\tilde{b} \leftrightarrows r\tilde{r}$$

These transitions lead to three linear combinations (linear superpositions), one of which, namely,

$$(y\tilde{y} + b\tilde{b} + r\tilde{r})/\sqrt{3},$$

is completely symmetric in color. It does not even have a hidden color charge, being totally colorless or, as is often said, white. The other two diagonal combinations can be presented in three physically equivalent ways:

$$(y\tilde{y} - b\tilde{b})/\sqrt{2} \quad \text{and} \quad (y\tilde{y} + b\tilde{b} - 2r\tilde{r})/\sqrt{6},$$

or

$$(b\tilde{b} - r\tilde{r})/\sqrt{2} \quad \text{and} \quad (b\tilde{b} + r\tilde{r} - 2y\tilde{y})/\sqrt{6},$$

or

$$(r\tilde{r} - y\tilde{y})/\sqrt{2} \quad \text{and} \quad (r\tilde{r} + y\tilde{y} - 2b\tilde{b})/\sqrt{6}.$$

Obviously, these combinations are obtained from one another by the cyclic replacement $y \to b \to r \to y$. A discussion of the coefficients in these linear superpositions and of the physical equivalence of the three distinct choices of diagonal superpositions is outside the scope of this book. What is important for us here is that each of the eight colored combinations (six explicitly colored and two with hidden color) corresponds to a gluon, so that there are eight gluons ($8 = 3 \times 3 - 1$).

It is essential that no direction in color space is favored: the three colored quarks play equivalent roles, and so do the three colored antiquarks; the same is true for the eight colored gluons. Color symmetry is absolute.

Quarks emit and absorb gluons and thereby exert the strong interaction on each other.

Let us consider, say, a red quark. When it emits the $r\tilde{y}$ gluon, it transforms (by virtue of the conservation of color) into a yellow quark because, by the rules of the game, the emission of anticolor is equivalent to the absorption of color. The emission of the $r\tilde{b}$ gluon by a red quark transforms it into a blue quark. Clearly, the same results are produced if the red quark absorbs the $\tilde{r}y$ gluon and the $\tilde{r}b$ gluon. In the first case, the quark becomes yellow and, in the second, blue. These processes of emission and absorption of gluons by a red quark can be written in the form

$$q_r \rightarrow q_y + g_{r\tilde{y}},\ q_r + g_{\tilde{r}y} \rightarrow q_y,$$
$$q_r \rightarrow q_b + g_{r\tilde{b}},\ q_r + g_{\tilde{r}b} \rightarrow q_b,$$

where q_r, q_y, q_b denote the red, yellow, and blue quarks of any flavor, $g_{r\tilde{y}}$, $g_{\tilde{r}y}$, $g_{r\tilde{b}}$, and $g_{\tilde{r}b}$ denote the red-antiyellow, antired-yellow, red-antiblue, and antired-blue gluons, respectively.

The emission and absorption of off-diagonal gluons by the yellow and blue quarks can be analyzed in a similar manner.

Obviously, the emission and absorption of diagonal gluons leaves the color of a quark unaltered.

The fact that gluons carry color charges results in a radical difference between gluons and photons. The photon has no electric charge and cannot, therefore, emit photons. On the contrary, gluons possess color charges, so a gluon can emit gluons. The smaller the mass of a charged particle, the easier it is for it to emit radiation. Gluons are massless, so the emission of gluons by gluons, if they were free, would be catastrophic. However, there is *no* catastrophe: strong interactions between gluons result in the confinement of both gluons and quarks. The strong interactions between color charges at distances of about 10^{-15} m become so strong that the individual color charges cannot break away and depart to larger distances. This means that the only combinations of color charges that can move freely are those with no net color charge.

Electrodynamics permits the existence of both isolated electrically-neutral atoms and isolated electrically-charged electrons and ions. However, in chromodynamics, only colorless or "white" hadrons, in which all colors are mixed in equal amounts, can exist in an isolated state. Thus, the π^+ meson spends equal time in each of the three permitted color states $u_y\tilde{d}_{\tilde{y}}$, $u_b\tilde{d}_{\tilde{b}}$, and $u_r\tilde{d}_{\tilde{r}}$; it is the sum (a superposition) of these states. The last statement, as well as the statement about gluons with hidden color, is unlikely to be quite clear to the unprepared reader. I regret this, but I have already remarked that certain things in elementary-particle physics do not have simple and demonstrative explanations.

■ It is appropriate at this point to say a few words about popular science literature. There is no doubt that popular science books and articles do help the lay reader, at least to some degree, in the vast, multidimensional, tangled maze of science. To that extent, this literature does much good. At the same time, some damage is also done. This literature provides descriptive, very approximate, comic-strip-type pictures of scientific theories and experiments (quite often, no other description is achievable in popular science books), and is likely to push the reader to a deceptive feeling of simplicity and complete understanding.

It is not unusual for the science comic-strip reader to gain the impression that scientific theories are largely, if not completely, arbitrary: "Lots of such explanations can be devised. Can't they?" In fact, the popular science literature is responsible for the steady flux of letters conveying illiterate "refutations" and "radical improvements" of relativity, quantum mechanics, and elementary-particle theory that floods many leading physics institutes.

The author of a popular-science book must provide not only a simple explanation of the simple but also a warning about the abstruse that only a specialist comprehends. ■

Colored quarks and gluons are not the inventions of the idle mind. Quantum chromodynamics is forced upon us by nature, and it has been verified, and continues to be verified, by an enormous amount of experimental evidence. It is one of the most complicated physical theories with extremely subtle mathematical apparatus that has not, as yet, been fully elaborated.

As far as we know, there is not a single fact that contradicts quantum chromodynamics. However, there are a number of phenomena for which this theory provides a qualitative rather than quantitative interpretation. Thus, our understanding of the mechanism by which hadron jets develop from quark + antiquark pairs created at short distances is still incomplete. We still lack a theory of confinement. These problems are being tackled by leading theoretical physicists all over the world. This work relies not only on traditional methods, i.e., on paper-and-pencil technology, but also on powerful modern computers. Continuous space and time are replaced in these "numerical experiments" by four-dimensional lattices containing about 10^4 sites, and gluon fields are analyzed on these lattices.

Leptons

The last few sections were devoted to the properties and structure of hadrons – the numerous relatives of the proton. We now turn to the relatives of the electron. The family name is leptons (from the Greek, *leptos*, meaning small; see p. 32). Like the electron, none of the other leptons participates in strong interactions, and they each have spin $\frac{1}{2}$. All leptons are treated, at our current level of knowledge, as truly elementary particles because none of them seems to have a structure like that found in hadrons. In this sense, leptons are said to be point-like particles.

The existence of the following leptons has now been established: three charged leptons e^-, μ^-, τ^-, and three neutral leptons ν_e, ν_μ, ν_τ (the latter three are called the electron neutrino, the muon neutrino, and the tau neutrino). Of course, each of the six leptons has its antilepton: e^+, μ^+, τ^+, $\tilde{\nu}_e$, $\tilde{\nu}_\mu$, and $\tilde{\nu}_\tau$.

We shall now briefly discuss each lepton separately.

The electron has already been discussed in detail earlier in the book.

The muon was discovered in cosmic rays. The process of discovery of the muon (from the first observation to the realization that this particle was a product of the charged-pion decay: $\pi^+ \to \mu^+\nu_\mu$, $\pi^- \to \mu^-\tilde{\nu}_\mu$) lasted for a decade: from the end of the 1930s to the end of the 1940s (the fact that the muon had its own neutrino – the so-called muon neutrino – was established much later, at the beginning of the 1960s).

The τ lepton was discovered in 1975 in the $e^+e^- \to \tau^+\tau^-$ reaction.

The masses of the muon and the τ lepton are 106 MeV and 1784 MeV, respectively. In contrast to the electron, the muon and the τ lepton are unstable. The lifetime of the muon is $2 \cdot 10^{-6}$ s and that of the τ lepton is $3 \cdot 10^{-13}$ s.

The muon decays into a single channel. The μ^- decays into $e^-\tilde{\nu}_e\nu_\mu$, and the μ^+ decays into $e^+\nu_e\tilde{\nu}_\mu$. The τ lepton has many decay channels:

$$\tau^- \to e^-\tilde{\nu}_e\nu_\tau,\ \tau^- \to \mu^-\tilde{\nu}_\mu\nu_\tau,\ \tau^- \to \nu_\tau + \text{mesons}$$
$$\tau^+ \to e^+\nu_e\tilde{\nu}_\tau,\ \tau^+ \to \mu^+\nu_\mu\tilde{\nu}_\tau,\ \tau^+ \to \tilde{\nu}_\tau + \text{mesons.}$$

This abundance of decay channels is due to the large mass of the τ lepton: it can decay into heavy particles that cannot appear among the muon decay products because of the conservation of energy.

Our information about the neutrinos is very meagre. The least-known is the ν_τ. We do not even know whether the mass of the ν_τ is zero or fairly large. The experimental upper limit is $m_{\nu_\tau} < 70$ MeV. The muon neutrino has a lower (but still quite large) upper limit: $m_{\nu_\mu} < 0.25$ MeV. The upper limit of the electron neutrino mass has been measured with much greater precision. One group of experimentalists has reported, at the limit of this precision, that $m \simeq 30$ eV. At present, physicists are waiting for independent tests of this conclusion in other laboratories.

Experiment shows that each of the charged leptons participates in weak interactions together with its neutrino: e with ν_e, μ with ν_μ, τ with ν_τ. For example,

$$n \to pe^-\tilde{\nu}_e, \qquad \pi^+ \to \mu^+\nu_\mu, \qquad \tau^+ \to \tilde{\nu}\, e^+\nu_e.$$

Generations of leptons and quarks

There are striking differences between quarks and leptons. The former are colored and have fractional charges, whilst the latter are colorless and have integral charges. However, they also have some common features: both are spin-$\frac{1}{2}$ particles and appear to be point-like particles at the present level of knowledge. This explains why leptons and quarks are referred to as fundamental fermions.

Fundamental fermions naturally divide into three groups, known as *generations*:

$$\begin{array}{ccc} u & c & t\,(?) \\ d & s & b \\ \nu_e & \nu_\mu & \nu_\tau \\ e^- & \mu^- & \tau^- \end{array}$$

The question mark against the t quark reminds us that, so far, this quark remains undiscovered. However, the fact that all entries in the first two generations (columns) are filled is a hint that the third generation has the same structure.

Particles of the first generation are the lightest, and those of the third generation are the heaviest.

The charged particles of the first generation are the blocks of which atoms are built and, although the electron neutrino tends to hide from view, its role is also important: without it, the Sun and stars would stop shining. In fact, the whole of the Universe rests on the shoulders of the first-generation particles.

We do not yet know, although we are beginning to guess, what purpose is served by particles in the other two generations. The longest-lived among them, the muon, lives for two microseconds ($2 \cdot 10^{-6}$ s); strange particles live for $10^{-8} - 10^{-10}$ s, and the rest of them live for less than 10^{-12} s. Having been produced with considerable difficulty in specially built accelerators, these particles perish practically at the moment of birth. The exceptions to this are the ν_μ and, possibly, the ν_τ, if it is massless or very light.

A number of questions inevitably arise: why study these ephemeral and exotic creations at all if they play no part in our life? Is the expenditure on costly accelerator laboratories justifiable?

I shall attempt, at the end of the book, to bring together different answers to the first question and lead the reader to an affirmative answer to the second. Here, I want to emphasize the following points.

First, the study of the strange, charmed, and other second- and third-generation particles has led to the discovery of the quark structure of ordinary nucleons, which consist of first-generation particles. Indeed, physicists were led to the idea of quarks by experimental investigations of strange particles, and the final confirmation of the existence of quarks was provided by charmonium. We cannot, therefore, rule out the possibility that further experiments on still more powerful accelerators will bring to light the internal structure of leptons and quarks themselves.

Second, studies of processes involving second- and third-generation fermions have enabled us to understand the fundamental forces that act not only between them, but also between fundamental fermions of the first generation. Without the numerous experiments of the last decade, and especially the collider experiments, it would have been hard to get conclusive evidence for the existence of gluons, which hold quarks together in the nuclei of ordinary atoms. Another example is the nature of the fundamental weak force, which was examined so thoroughly by analyzing the weak interaction between leptons and quarks of all three generations. The culmination of this program was the discovery in 1983 of the so-called intermediate vector bosons, namely, W^+, W^-, and Z^0, which are the fundamental carriers of the weak interaction.

Decays of leptons and quarks

Quarks q and antiquarks $\tilde{q}$ of a given flavor are produced in strong and electromagnetic interactions only in $q\tilde{q}$ pairs. Consequently, were it not for the weak interaction, the flavors of quarks and leptons would be conserved and all twelve fundamental fermions would be stable.

Examples of the weak processes, such as the β-decay of the neutron and the related weak fusion reaction fuelled by the protons and electrons in the Sun, have already been introduced. Let us now consider how these and other weak processes can be described in terms of quarks. We begin with the β-decay of the neutron. We recall that p = uud and n = ddu. The β-decay of the neutron ($n \to pe^-\tilde{\nu}_e$) then reduces to the β-decay of the d quark (Figure 12):

$d \to ue^-\tilde{\nu}_e$.

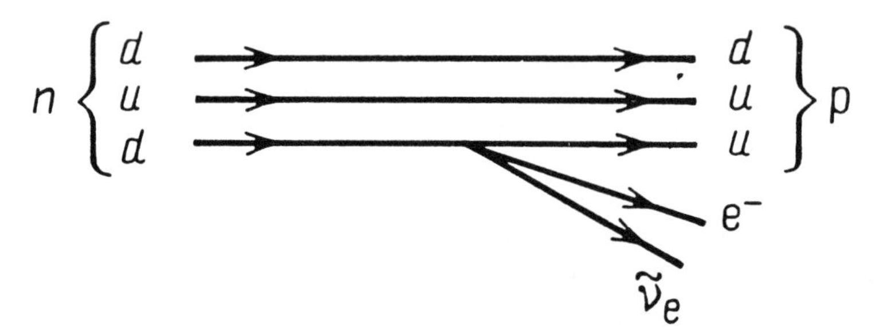

Figure 12. Quark diagram of the β-decay of the neutron.

Likewise, it is readily verified that the solar reaction $pe^-p \to D\nu_e$ (we recall that D stands for the deuteron, the nucleus of heavy hydrogen) reduces to the reaction

$ue^- \to d\nu_e$.

This last process can be deduced from the decay of the d quark by two operations: (1) replace one of the antiparticles (in this case, $\tilde{\nu}_e$) in the final state with its counterpart (in this case, ν_e) in the initial state and (2) reverse the time arrow, i.e., exchange the final and initial states.

The meaning of the second operation is clear. It can be called time reversal because, as in a movie projected backward, the process takes place in the reverse direction.

The legitimacy of the first operation is less obvious. We must dwell on it here because it plays a key role in the theoretical analysis of all processes in which elementary particles participate.

However complex a natural process may be, we may be sure that conservation of electric, baryonic, leptonic, and other possible charges will allow processes in which the emission (absorption) of any of the particles is replaced with the absorption (emission) of the corresponding antiparticle. The "algebra of antiparticles" (transposition of particles from one side of the equation describing the reaction to the other side, and the simultaneous reversal of the sign of the charges) is based on the fact that all the charges of a particle are opposite to those of the corresponding antiparticle. This "algebra" is the foundation of the modern theory of elementary particles, i.e., quantum field theory. As an exercise, try to verify that all electromagnetic processes, as well as all gluon-quark processes discussed on the preceding pages, obey this antiparticle algebra.

However, the conservation of all charges is not enough to make a process occur in nature. It is also necessary (and sufficient)[⊆] that the process satisfies the laws of conservation of physical quantities such as energy and momentum. And here, everything depends on the specific conditions in which the interacting particles are found. For example, the reaction $e^-p \rightarrow n\nu_e$ is forbidden (and we must be grateful to nature for this) for the electron and proton forming the hydrogen atom because the neutron is heavier than the proton and electron taken together. Just imagine what the world would be like if the hydrogen lifetime were only a few seconds long!

On the other hand, the reaction $e^-p \rightarrow n\nu_e$ does take place in some atoms. One example is the unstable beryllium isotope ${}^7_4Be$. The nucleus of this atom captures one of the atomic electrons, and the result is the lithium isotope ${}^7_3Li$. This is possible because the atom with the ${}^7_3Li$ nucleus is lighter than the atom with the ${}^7_4Be$ nucleus. A similar situation occurs in the nuclear fusion reaction $pep \rightarrow D\nu_e$: the deuteron mass is less than the sum of masses of two protons and one electron.

It is now evident that, if a hydrogen target is bombarded with a beam of electrons of energy greater than the threshold for the reaction

$$e^-p \rightarrow n\nu_e,$$

the reaction is allowed. The same is true for the reaction

$$\tilde{\nu}_e p \rightarrow e^+n$$

if the energy of $\tilde{\nu}_e$ exceeds the threshold $(m_n - m_p + m_e)c^2$. The reaction has actually been observed in low-energy antineutrino beams on nuclear reactors and in high-energy antineutrino beams on accelerators. As for the reaction

---

[⊆]In physics, anything that is not forbidden has to take place. A consistent application of this principle should enable you to understand and arrange in order much of what happens in the kaleidoscopic world of reactions and decays of elementary particles. Note that it is this principle that demands that an atom in the excited state must return to the ground state.

$\nu_e n \to e^- p$,

it was observed not on free neutrons (there are no sufficiently dense targets consisting of free neutrons), but on neutrons bound in the deuteron and heavier nuclei.

By manipulating time reversal and the algebra of antiparticles, we can readily write out numerous quark processes related to the quark decay $d \to ue^-\tilde{\nu}_e$. An example is the transformation of a quark-antiquark pair into a lepton-antilepton pair:

$d\tilde{u} \to e^-\tilde{\nu}_e$ or $u\tilde{d} \to e^+\nu_e$, and so on.

The first of these processes is responsible for the decay $\pi^- \to e^-\tilde{\nu}_e$, and the second is responsible for the decay $\pi^+ \to e^+\nu_e$ (Figures 13 and 14).

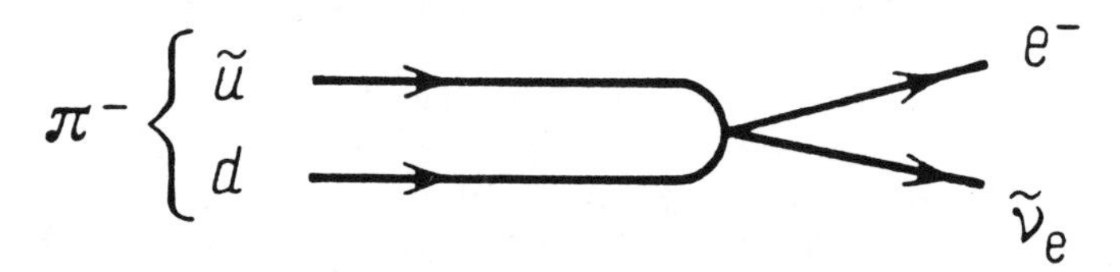

Figure 13. Quark diagram of the decay $\pi^- \to e^-\tilde{\nu}_e$.

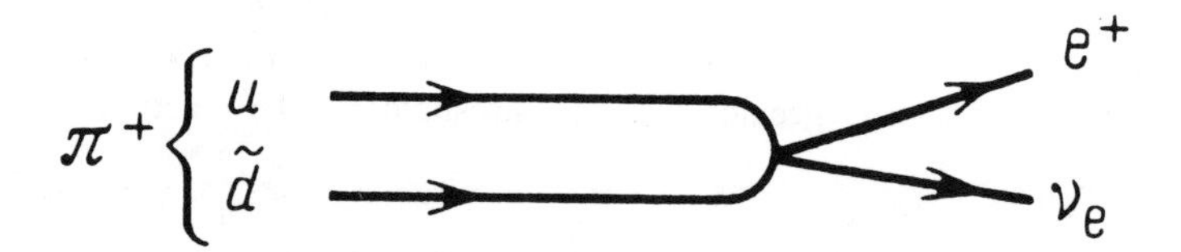

Figure 14. Quark diagram of the decay $\pi^+ \to e^+\nu_e$.

We thus find that the weak interaction of the quark pair ud with the lepton pair $\nu_e e$ is the basis for a whole class of processes. The quark pair ud interacts with two other lepton pairs, $\nu_\mu\mu$ and $\nu_\tau\tau$, in exactly the same way. For example, consider the following decays (Figures 15 and 16):

$\pi^+ \to \mu^+\nu_\mu$, $\tau^+ \to \pi^+\tilde{\nu}_\tau$

and the capture of a muon in hydrogen (Figure 17):

$\mu^- p \to n\nu_\mu$.

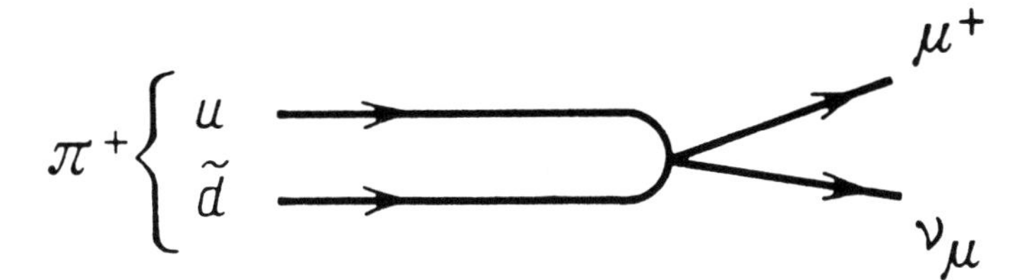

Figure 15. Quark diagram of the decay $\pi^+ \rightarrow \mu^+\nu_\mu$.

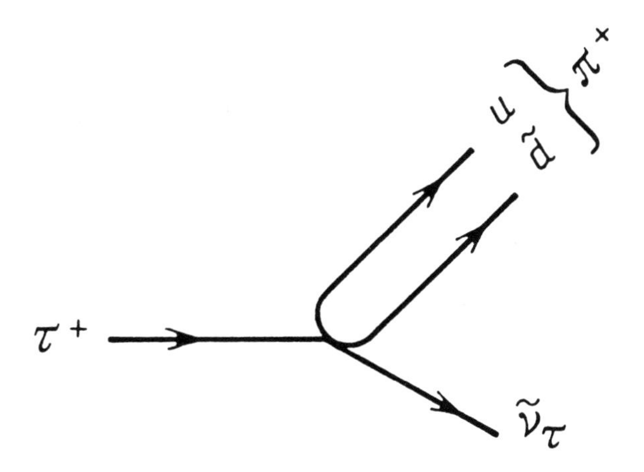

Figure 16. Quark diagram of the decay $\tau^+ \rightarrow \pi^+\tilde{\nu}_\tau$.

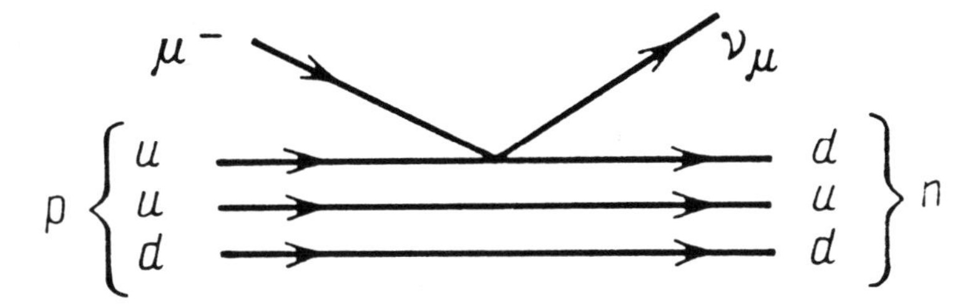

Figure 17. Quark diagram of muon capture: $\mu^- p \rightarrow n\nu_\mu$.

Note that each charged lepton is "loyal" to its neutrino.

Turning now to other quark pairs, it would seem natural to expect that quarks, like leptons, undergo transitions only within the generations to which they belong. If this were true, the weak interaction would involve only the three quark pairs ud, cs, and tb. The s quark and b quarks would then be stable because the s quark is much lighter than the c quark, and the b quark is much lighter than the t quark. This would mean that strange particles and hadrons containing b quarks would also be stable. We know that this is not the case and that there are numerous weak processes in which quarks of one generation transform into quarks of other generations (Figure 18). The solid lines in Figure 18 indicate leptonic and quark "weak pairs", observed experimentally, and the dashed lines indicate pairs that are expected but have not yet been observed.

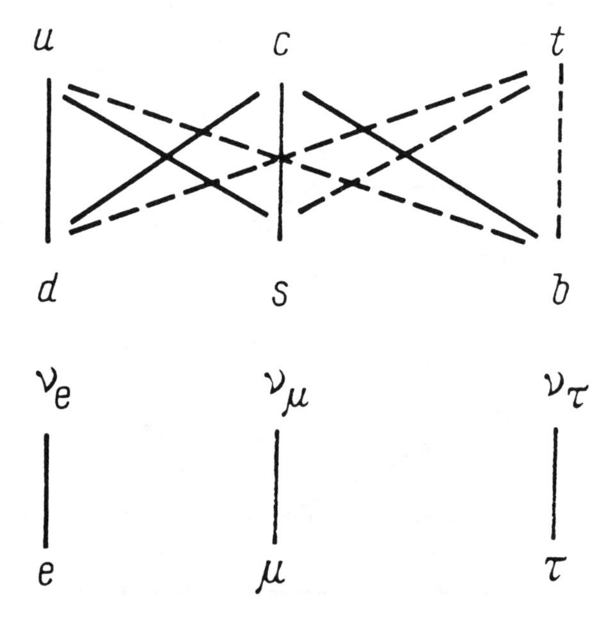

Figure 18. Three generations of leptons and quarks. Connecting lines show three "weak pairs" of leptons and nine "weak pairs" of quarks.

All in all, three lepton pairs and nine quark pairs must participate in the weak interaction. Interactions between lepton and quark pairs yield numerous lepton-hadron reactions and decays. Here are some additional examples of decays involving both leptons and hadrons (sometimes, these decays are said to be semileptonic):

$K^+ \rightarrow \mu^+\nu_\mu$, $\Lambda \rightarrow pe^-\tilde{\nu}_e$, $F^+ \rightarrow \tau^+\nu_\tau$,
$\tau^+ \rightarrow K^+\pi^0\tilde{\nu}_\tau$, and so on.

In addition to lepton-hadron processes, there must exist, and do exist, both purely leptonic processes (Figure 19) and purely hadronic (so-called nonleptonic) processes. The latter are caused by weak interactions between quark pairs, for example,

(ud) (du) or (ud) (cs) or (cs) (bc).

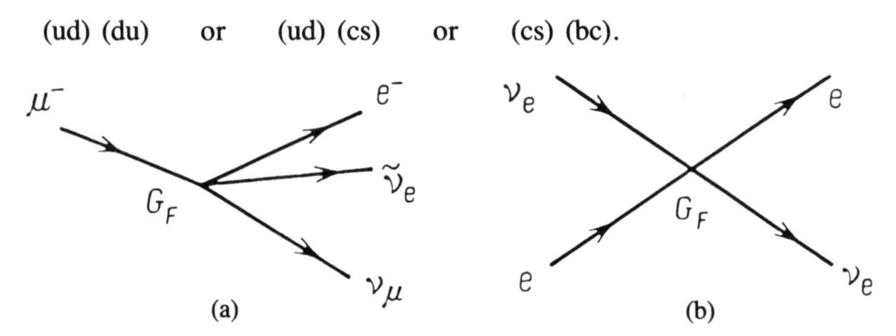

Figure 19. Two examples of purely leptonic weak processes: (a) muon decay and (b) scattering of an electron neutrino by an electron. Each process involves four fermions whose interaction is characterized by the Fermi constant G_F.

Examples of nonleptonic decays due to the (du) (us) interaction are the decays (Figure 20)

$K^+ \rightarrow \pi^+\pi^0$, $\Lambda \rightarrow p\pi^-$.

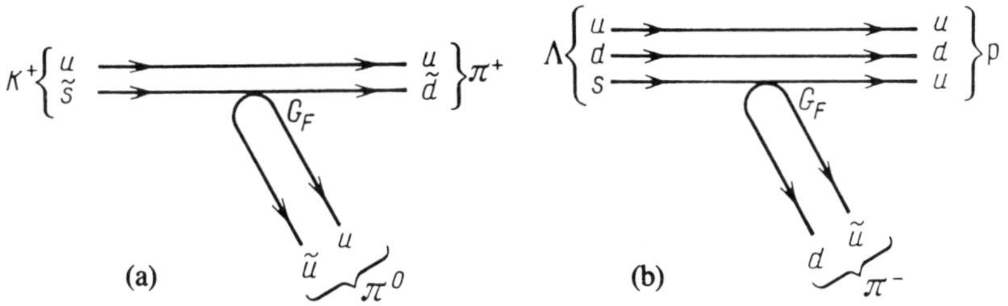

Figure 20. Two examples of quark diagrams describing purely hadronic weak processes: (a) $K^+ \rightarrow \pi^+\pi^0$ decay and (b) the $\Lambda \rightarrow p\pi^-$ decay. The weak four-fermion interaction characterized by the Fermi constant G_F takes place between quarks.

It can be demonstrated that 12 "weak pairs" must give $\frac{1}{2}12\times(12 + 1) = 78$ different interactions, each of which generates, like the (ud) $(e\nu_e)$ interaction, a host of different processes.

So far, most of the 78 "bare" interactions have escaped observation: 33 involve the yet undiscovered t quark; and, in some cases, suitable targets or beams are not available, as for $(\nu_\tau\tau)$ $(\tau\nu_\tau)$. Nevertheless, whenever an interaction was predicted and its observation feasible, it was invariably observed.

The diversity of weak processes is "controlled" by a single fundamental constant, namely, the so-called four-fermion interaction constant G_F (also known as the Fermi constant). In SI units, $G_F = 1.4\cdot 10^{-62}$ J.m^3, but it is more instructive to convert G_F into electron volts. By separating explicitly the factors $\hbar$ and c, you will find that (please verify this):

$$G_F \simeq 1.2\cdot 10^{-5}\hbar^3c^3\ \mathrm{GeV}^{-2}.$$

The probabilities of all weak reactions and weak decays are proportional to G_F^2. For example, the muon lifetime τ_μ is given in terms of G_F and muon mass m by the following formula:

$$1/\tau_\mu = G_F^2 m^5 c^4/192\pi^3\hbar^7.$$

This suggests another exercise: please verify that $\tau_\mu = 2 \times 10^{-6}$ s.

Faster decays correspond to greater energy release. For example, the lifetimes of particles heavier than the muon, such as the τ lepton, c quark, and b quark, are shorter than the muon lifetime by six or seven orders of magnitude. On the other hand, the neutron lives a billion times longer than the muon. Roughly speaking, the probability of weak decays increases as the fifth power of energy release. In the decay of the τ lepton, c quark, or b quark, practically the entire rest energy is converted into the kinetic energy of the decay products, and the energy released is high. In neutron decay, the energy released is only 0.8 MeV – hence, the long life of the neutron.

Virtual particles

Why is it that each "weak pair" interacts with every other such pair? An analogy with electrodynamics will again (as on many previous occasions) help us to find the answer.

Why is an electron scattered by a proton? Because each of them interacts with the electromagnetic field, i.e., with photons. The scattering of a free electron by a free proton is accompanied by energy and momentum transfer and can be described in the following manner. One of the particles, say, the electron, emits a photon and the other (the proton) absorbs it. However, the photon exchanged by the electron and the proton is not an ordinary free photon. Indeed, consider the emission of a photon by an electron that was free before the emission and remains free after it. By applying energy and momentum conservation to this elementary "sub-process", we can readily ascertain that the energy of the emitted photon is not equal to its momentum (times the velocity of light c), as expected for real free photons. Were it so, a free electron at rest would constantly emit photons, in absolute contradiction to the conservation of energy.

We recall that a free particle of mass m must obey the relation $E^2 - \mathbf{p}^2c^2 = m^2c^4$, whilst $m = 0$ for a photon. However, when the photon is exchanged in electron-proton scattering, we have $E^2 - \mathbf{p}^2c^2 < 0$. This is particularly easy to demonstrate if we consider ep scattering in the center-of-mass frame of the electron-proton system, in which the energy of the electron remains unaltered and only the direction of its momentum changes. Therefore, the photon exchanged by the electron and proton transfers momentum $\mathbf{p}$, but its energy E is zero, so that $E^2 - \mathbf{p}^2c^2 = -\mathbf{p}^2c^2 < 0$.

Photons satisfying the condition

$$E^2 - \mathbf{p}^2c^2 \neq 0$$

are called *virtual photons*. In this context, the word "virtual" is opposite to the word "real". The distance to which a virtual photon can depart from its source is short and decreases with increasing virtuality of the photon. When an electron passes a proton and absorbs a virtual photon, scattering takes place, and both the electron and proton change their momenta. By definition, there are no events in which a virtual proton is emitted but not absorbed.

The concept of the virtual particle, i.e., particle for which

$$m^2c^4 \neq E^2 - \mathbf{p}^2c^2,$$

plays an important role in quantum field theory. In fact, all processes, even the most complex, can be reduced to the emission, propagation, and absorption of real and virtual particles.

The so-called Feynman diagrams, or graphs, are very convenient in the

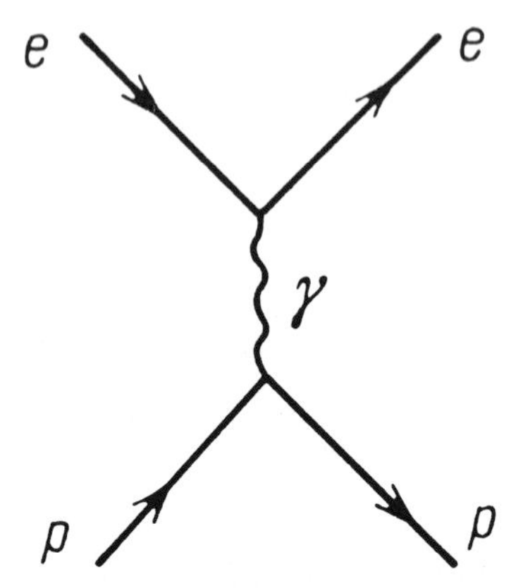

Figure 21. Feynman diagram describing the elastic scattering of an electron by a proton. The wavy line represents a virtual photon.

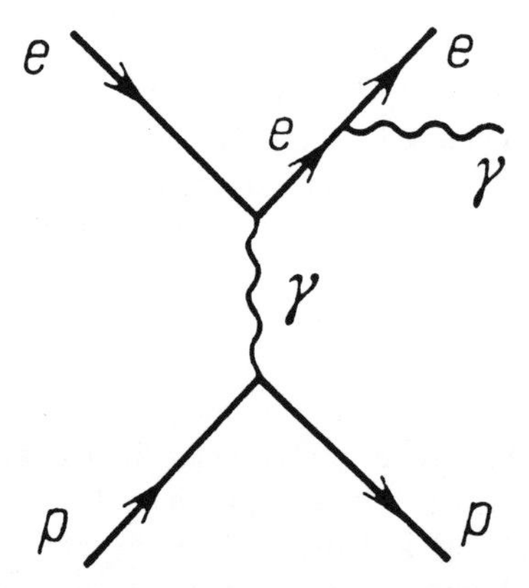

Figure 22. Feynman diagram describing the inelastic scattering of an electron by a proton, in which a bremsstrahlung photon is emitted. This diagram contains two virtual particles: a virtual photon and a virtual electron.

theoretical analysis of all these processes. Real particles are represented in this approach by "rays", i.e., lines arriving from infinity or departing to infinity, whilst the propagation of virtual particles is shown by "segments", i.e., by lines joining other lines. Usually (although not necessarily), fermions are indicated by straight lines and bosons by wavy lines. For example, Figure 21 represents the scattering of an electron by a proton through virtual photon exchange. Figure 22 represents the scattering of an electron by a proton, in which the electron additionally emits a real *bremsstrahlung photon* (from the German *die Bremse* meaning brake, and *die Strahlung*, meaning emission; the electron momentum diminishes so that the electron slows down). The electron line connecting the points of emission of the virtual and real photons in this figure represents a virtual electron (we already know that a free electron cannot emit a free photon and remain free all the time).

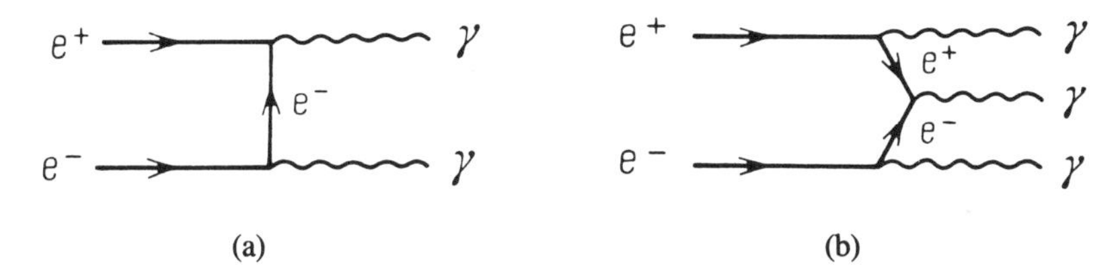

Figure 23. Feynman diagrams describing the annihilation of an electron and a positron into two photons (a) and into three photons (b).

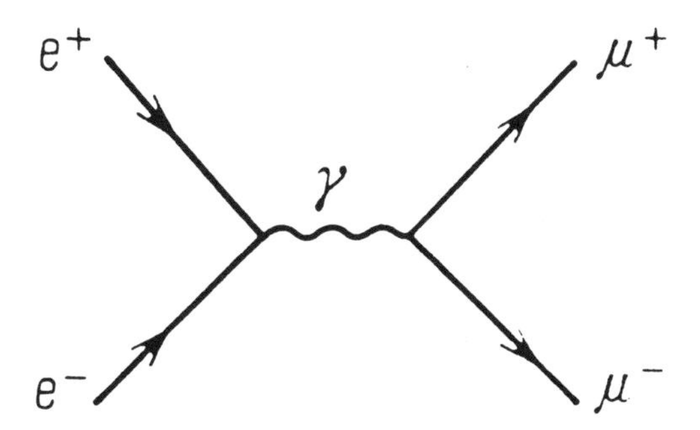

Figure 24. Feynman diagram describing the annihilation of an electron and a positron into a pair of muons.

The Feynman diagrams of Figures 23 and 24 show some electron annihilation processes. Figures 23a and b represent the annihilation of a positron and an electron into two and three photons, respectively. Here, the virtual particles are the electrons. Figure 24 shows the annihilation of an electron and a positron into a $\mu^+\mu^-$ pair. When this process is analyzed in the center-of-mass reference frame, in which the sum of the momenta of the electron and positron is zero, the virtual photon has no momentum, but does have energy.

The points at which lines join other lines in Feynman diagrams are called *vertices*. Vertices with their adjacent lines are called *vertex parts*. They describe the emission and absorption of particles, while the lines describe their propagation. Each vertex is characterized by a certain quantity (not necessarily dimensionless), called the interaction constant or *coupling constant*. The smaller this coupling constant, the weaker is the corresponding interaction.

We already know that the coupling constant for weak decays and reactions is the Fermi constant (for example, see Figure 19).

The coupling constant for electromagnetic processes is the electric charge e. This charge is also commonly characterized (see p. 20) by the dimensionless ratio $\alpha = e^2/\hbar c = 1/137$, which, ever since the early days of the theory of atomic spectra, has been called the *fine structure constant*. (Fine structure is the splitting of atomic levels by the interaction between the magnetic moment of the electron, which is associated with its spin, and the magnetic moment associated with its orbital motion.)

The potential energy of Coulomb attraction between an electron and a proton, due to the virtual-photon exchange shown in Figure 21, has the form

$$U(r) = -\frac{e^2}{r} = -\frac{\alpha\hbar c}{r}.$$

The slow fall in potential energy with increasing r (proportional to 1/r) occurs because the photon is massless. It can be shown that, if a photon had a mass m, the potential corresponding to the exchange of this "massive photon" would be

$$U(r) = -\frac{\alpha\hbar c}{r}\exp\left(-\frac{r}{\hbar/mc}\right),$$

i.e., it would decrease exponentially for $r > \hbar/mc$. This potential bears the name of the Japanese physicist Yukawa. We recall that $\hbar/mc$ is the Compton wavelength of a particle of mass m.

The Yukawa potential rapidly becomes less effective as the mass of the "massive photon" increases. This is not unexpected. The greater the photon mass, the smaller the space "allowed" for it by the quantum-mechanical uncertainty relation, and the smaller is the effect of the photon on the electron cloud of the atom. Virtual particles are often compared in popular physics books with tennis balls, the exchange of which keeps the players within the court. Of course, this is a very superficial comparison because the tennis ball does not convey physical attraction; but, if we stretch the point a little, it becomes clear that the heavier the ball, the smaller the effective size of the tennis court. A 100-kg ''ball'' cannot be thrown far.

Feynman diagrams contain the recipe for calculating the probabilities of the processes involved. In quantum mechanics, the probability of a process is given by the square of the modulus of a certain complex quantity, called its *amplitude*. Feynman diagrams prescribe the algorithm for computing the amplitude. Each element of the diagram corresponds to a certain factor, so that lines representing real particles correspond to the wave functions of these particles, whilst vertices correspond to the coupling constants. Lines representing virtual particles in Feynman diagrams correspond to propagation functions of these virtual particles, called *propagators*.

The propagator of a virtual photon with four-dimensional momentum q is

$$-(1/q^2),$$

where $q^2 = q_0^2 - \mathbf{q}^2$, $q_0 = E/c$, E is the energy of the photon, and $\mathbf{q}$ is its momentum. Since the probability of the process is proportional to the square of the amplitude, we may conclude that the probability that momentum q will be transferred in, say, the scattering of an electron by an electron will fall as $1/q^4$.

The propagator of a virtual vector boson of mass m (''heavy photon'') with four-dimensional momentum q is

$$\frac{1}{(m^2 - q^2)}$$

We note in passing that, when $m^2 >> |q^2|$, the propagator is practically independent of momentum and reduces to the constant $1/m^2$.

We have examined the exchange of a ''very heavy photon'' because we shall need it in connection with the weak interaction. In 1983, it was finally shown experimentally that the weak interaction owed its existence to the exchange of very heavy vector particles. Just as all electromagnetic interactions between leptons and quarks are conveyed by virtual photons, all weak processes involving leptons and quarks are conveyed by heavy intermediate bosons W^+, W^-, and Z^0. The imaginary massive photon (no such photon exists in nature) was considered above merely as an aid to our discussion of heavy intermediate bosons.

Currents

To give a reasonably clear exposition of the emission and absorption of intermediate bosons, we have to introduce one more concept, namely, that of the *current*. Actually, it has been implicit throughout our discussion, ever since the emission and absorption of photons and the interaction between an electron and the electromagnetic field were mentioned.

Color currents were involved when we were discussing the emission and absorption of gluons, whilst weak currents were implicit in the "algebra of antiparticles" in weak interactions.

Physics develops through the evolution of progressively more abstract concepts, which describe nature more and more adequately but become further removed from direct everyday experience. The concepts of *charge* (not necessarily electric charge) and *current* are among the most fundamental, primary concepts of modern physics. Their meaning cannot be explained fully without employing the mathematical language of quantum field theory. However, the reader can make some progress by learning how to operate with these concepts even without understanding them completely.

Let us begin with the electromagnetic current. As a crude first approximation, you may consider that the electromagnetic current is a moving electric charge, i.e., it is almost the same as ordinary electric current. What is important for us is that the electromagnetic current is a source of photons. In the Feynman diagram, the electromagnetic current of the electron corresponds to the portion of the electron-photon vertex part (Figure 25a) that contains only the electron lines (Figure 25b).

The electron electromagnetic current is responsible for the emission and absorption of photons not only by the electron, but also by the positron (Figure 25c). Moreover, it is responsible for the creation and annihilation of electron-positron pairs at the vertices of Figures 25d and 25e. In fact, the electromagnetic current is responsible for all electron-positron transitions and all interactions of electrons and positrons with real and virtual photons. The same is true for the electromagnetic currents of muons, tau

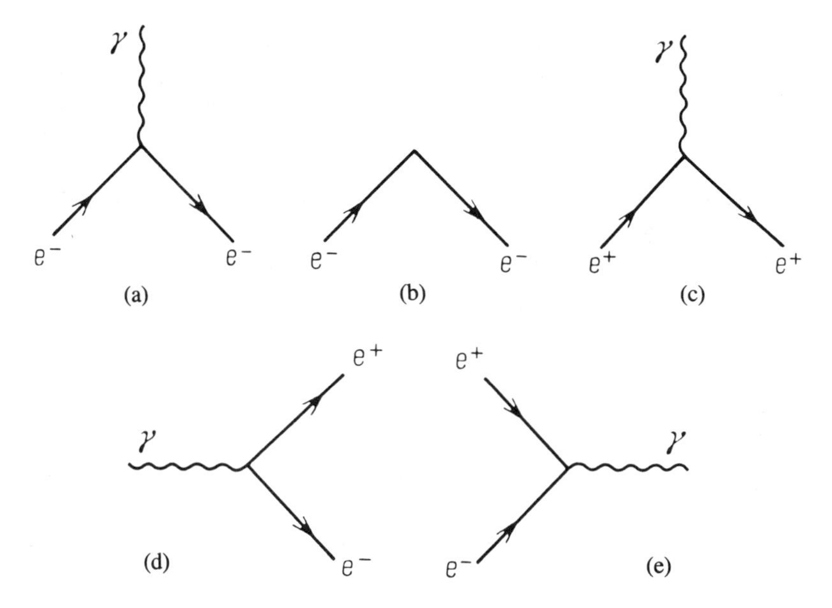

Figure 25. Parts of Feynman diagram describing (a) the emission and absorption of a photon by an electron and (c) by a positron, (d) the creation of an e^+e^- pair by a photon, and (e) the annihilation of an e^+e^- pair into a photon. The part of (a) that corresponds to the electromagnetic current is shown in (b).

leptons, quarks, and the corresponding antiparticles. All electromagnetic interactions of leptons and quarks can be described as the interaction between the photon and the total electromagnetic current j^{em}. The coupling constant of this interaction is the electron charge e. The electromagnetic current has the following form:

$$j^{em} = \bar{e}e + \bar{\mu}\mu + \bar{\tau}\tau + \tfrac{1}{3}\bar{d}d + \tfrac{1}{3}\bar{s}s + \tfrac{1}{3}\bar{b}b - \tfrac{2}{3}\bar{u}u - \tfrac{2}{3}\bar{c}c - \tfrac{2}{3}\bar{t}t,$$

where the fractional factors are the ratios of quark charges to the electron charge (the significance of the bar over some symbols will be explained below).

■ Actually, this is an oversimplified expression for the electromagnetic current, since j^{em} is a four-dimensional (i.e., four-component) vector,

$$j^{em}_{\mu} \ (\mu = 0, 1, 2, 3).$$

Only the time component j_0 is nonzero for an electron at rest, but all four components are nonzero for a moving electron.

Unfortunately, this remark is difficult to explain outside the content of quantum electrodynamics. However, the fact that the electromagnetic current is a four-dimensional vector is too important to be ignored. We recall that the photon is a vector particle. Vector particles are emitted and absorbed by vector currents. This will be discussed in greater detail later in this book. ■

We note that each quark current is the sum over three quark colors:

$$\bar{u}u = \bar{u}_y u_y + \bar{u}_b u_b + \bar{u}_r u_r,$$
$$\bar{c}c = \bar{c}_y c_y + \bar{c}_b c_b + \bar{c}_r c_r,$$

and so on.

The photon is "color-blind". All terms of the electromagnetic current are thus identical except for the factors $\frac{1}{3}$ and $-\frac{2}{3}$. We can therefore examine the physical meaning of current by confining our attention to one of the terms, say, $\bar{e}e$. The symbols e and $\bar{e}$ in this expression represent the so-called *operators* which, in quantum field theory, describe the creation and annihilation of particles and antiparticles. The operator e annihilates an electron, and $\bar{e}$ creates an electron. The current operator $\bar{e}e$ annihilates an existing electron and immediately creates a new electron, but not necessarily in the same state ("The King is dead. Long live the King!"). Actually, each of the operators e and $\bar{e}$ has not one but two responsibilities. In addition to annihilating electrons, the operator e also creates positrons, and the electron-creating operator $\bar{e}$ also annihilates positrons. This leads to the "algebra of antiparticles", described above. It also enables us to treat the positron, in the language of Feynman diagrams, as an electron moving "backward in time" (each of the virtual-electron lines in Figures 23a and 23b can be relabelled with e^+ instead of e^-, with the arrow reversed).

So far, we have discussed the electromagnetic current of leptons and quarks. Color and weak currents are organized in a similar manner.

The colored quark currents responsible for the emission of gluons by quarks have color indices that are determined by the colors of the particular gluon that is emitted or absorbed. Thus, the current $\bar{u}_y u_r$ transforms a red u quark into a yellow u quark by emitting a red-antiyellow ($r\tilde{y}$) gluon or absorbing a yellow-antired ($y\tilde{r}$) gluon. The same current describes the transformation of an antiyellow antiquark into an antired antiquark, the creation of a "yellow quark – antired antiquark" pair, and the annihilation of an "antiyellow antiquark – red quark" pair.

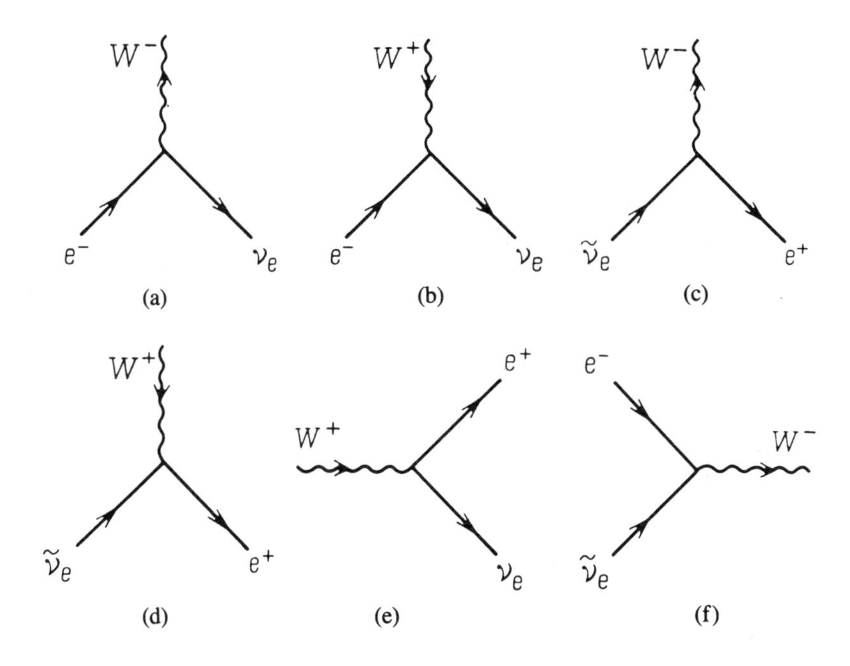

Figure 26. Vertex parts of Feynman diagrams describing the emission of a W^- boson and the absorption of a W^+ boson by the weak positive current $\tilde{\nu}_e e$.

We now turn to weak currents. Like electromagnetic currents, the weak currents of quarks are colorless, i.e., they are sums over three color indices. In fact, we have already introduced weak currents in the section *Decays of leptons and quarks* when we referred to "weak pairs" of leptons and "weak pairs" of quarks. Each such pair corresponds to two conjugate currents (we shall call them positive and negative currents). For example, the pair $e\nu_e$ corresponds to the currents $\tilde{\nu}_e e$ and $\bar{e}\nu_e$. The former describes the transformation of an electron into a neutrino (and of an antineutrino into a positron) via the absorption of a W^+ boson or emission of a W^- boson, and the transformation of a W^+ into an $e^+\nu_e$ pair, or of an $e^-\tilde{\nu}_e$ pair into a W^- (Figure 26). The second current describes the charge-conjugate processes (Figure 27).

The total positive current responsible for the absorption of positive W bosons consists of twelve terms: three lepton and nine quark terms. The same is true for the negative current. The phrases "positive current" and

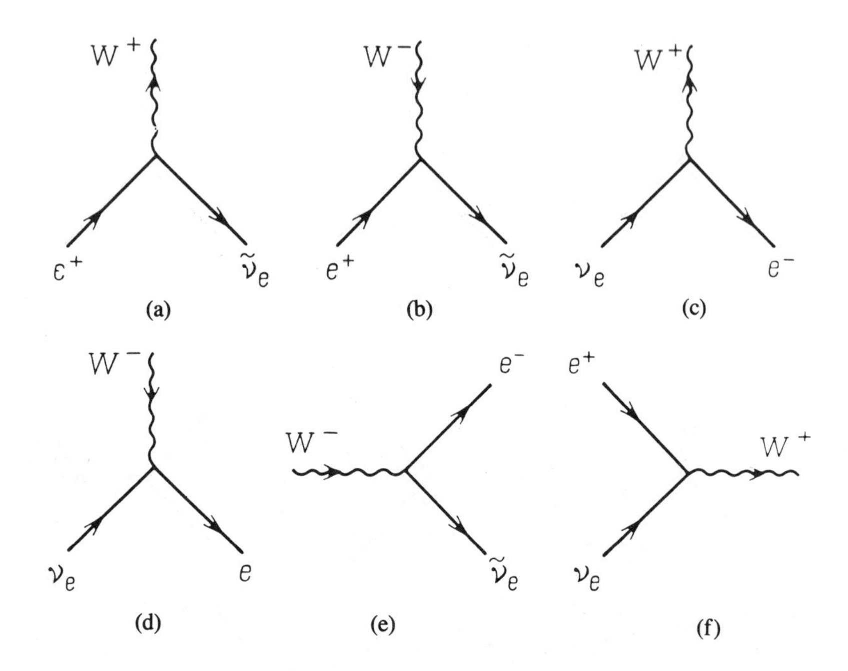

Figure 27. Vertex parts of Feynman diagrams describing the emission of a W^+ boson and the absorption of a W^- boson by a weak negative current $\bar{e}\nu_e$.

"negative current" are not actually standard terms in the literature: we have introduced them to make the presentation clearer. Normally, one refers to them as the charged current and its conjugate current. All weak processes that we discussed on the preceding pages occur because the charged current interacts with the conjugate current via W-boson exchange.

The differences between the lepton and quark currents, illustrated schematically in Figure 18, will be better understood by inspecting Figures 28 and 29.

The lepton current can be written as the sum of scalar products of the unit vectors shown in Figures 28a and 28b. The unit vector of each neutrino is parallel to the unit vector of the corresponding charged lepton and perpendicular to the other two unit vectors, so the lepton current consists of only three components.

The quark current can be written as the sum of scalar products of the

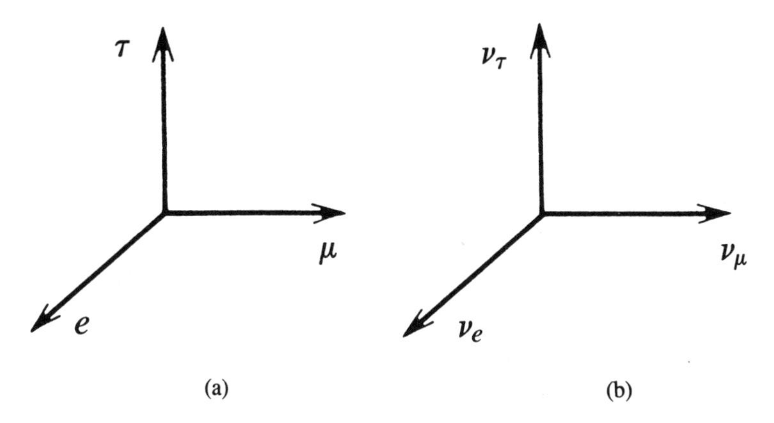

Figure 28. Three mutually orthogonal unit vectors corresponding to three charged leptons (a) and three neutrinos (b). Weak leptonic currents correspond to the scalar products of unit vectors of (a) and (b). Since the corresponding leptonic unit vectors are pairwise parallel, there are only three negative leptonic currents ($\bar{e}\nu_e$, $\bar{\mu}\nu_\mu$, $\bar{\tau}\nu_\tau$) and three positive currents.

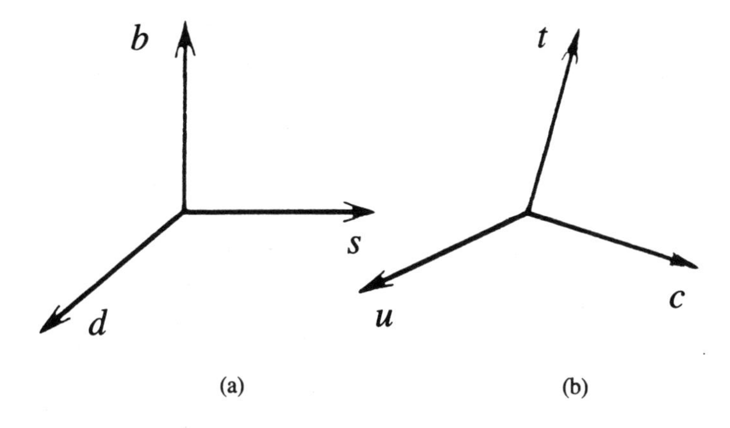

Figure 29. Three mutually orthogonal unit vectors corresponding to three quarks with charges $-\frac{1}{3}$ (a) and three quarks with charges $+\frac{2}{3}$ (b). Weak quark currents correspond to the scalar products of the unit vectors of (a) and (b). Since the triples of quark unit vectors (a) and (b) are rotated relative to each other, there are nine weak negative quark currents ($\bar{d}u$, $\bar{s}u$, $\bar{b}u$, $\bar{d}c$, $\bar{s}c$, $\bar{b}c$, $\bar{d}t$, $\bar{s}t$, $\bar{b}t$) and nine positive currents.

unit vectors shown in Figures 29a and 29b. The triple of quark unit vectors (u, c, t) is not parallel to the triple of quark unit vectors (d, s, b), so the quark current consists of nine components. However, the coefficients in front of these nine terms are smaller than those in lepton currents (the corresponding cosines are less than one). These coefficients determine the relative strengths of different quark-quark and quark-lepton interactions.

By comparing the probabilities of different weak processes, we can determine experimentally the angles between the unit vectors shown in Figures 29a and 29b. Experiments show that this orientation is approximately as shown in these figures: the u quark "prefers" to transform into the d quark rather than into the s or b quarks, and the c quark prefers to transform into the s quark, since the unit vector c is almost parallel to the unit vector s. This immediately tells us that the t quark (which has not been discovered, so far) will show a preference for the b quark in its weak interactions: the unit vectors t and b must be practically parallel.

C, P, and T symmetries

The weak interaction is actually considerably more powerful than its stronger companions, the electromagnetic and strong interactions. Only the weak interaction is capable of changing the flavors of quarks and leptons. Only the weak interaction is strong enough to violate the so-called discrete symmetries:

– symmetry under charge conjugation, i.e., under the replacement of all particles involved in a process by their antiparticles; this is known as *C symmetry* (C for charge)

– symmetry under spatial reflections, i.e., under the replacement of a process by the mirror-reflected process; this symmetry is *P symmetry* (P for parity)

– symmetry under time reversal, i.e., under the replacement of a process by the reverse process; this is known as *T symmetry* (T for time).

Before 1956, physicists regarded these three symmetries as being as immutable as the uniformity and isotropy of space and the uniformity of time. However, certain oddities in the decay of strange mesons led to the suspicion that this was not the case. Special experiments then revealed that the P and C symmetries were not only broken in all weak processes but that this occurred to the maximum possible extent.

This was discovered in a series of experiments in which the polarization of particles was measured. The word *polarization* has several meanings in physics. The meaning we shall need here is the orientation of the spins of an ensemble of particles. When the spins have arbitrary orientations, the polarization is said to be zero. When all of them are identically oriented, the polarization is said to equal unity.

The experiments of 1956–1957 showed that the particles created in weak processes were longitudinally polarized, i.e., they were polarized along their momenta. According to the common convention, a particle is said to be right-polarized if its spin points along its momentum and left-polarized if its spin points against its momentum.

It was found experimentally that the e^-, ν_e, μ^-, ν_μ created in weak processes were left-polarized, whilst the e^+, $\tilde{\nu}_e$, μ^+, $\tilde{\nu}_\mu$ were right-polarized,[⊆] the degree of polarization being v/c in both cases, where v is the velocity of the particle and c the velocity of light in vacuum. The fact that the longitudinal polarization tends to zero as the velocity v tends to zero should cause no surprise: by definition, there can be no longitudinal polarization in the limiting case of zero momentum.

The discovery of the longitudinal polarization of particles meant that the mirror-reflection symmetry had to be abandoned. The discovery that particles and antiparticles had different polarizations meant that the charge conjugation symmetry was also violated. The validity of the latter conclusion is self-evident, but the former needs additional clarification.

Consider a Cartesian system of coordinates in which the position vector **r** defines a point with coordinates x, y, z (Figure 30a). Now, take a mirror-reflected coordinate system (Figure 30b). The new coordinates of the same point are −x, −y, −z. In other words, the radius vector **r** changes sign under reflection of the coordinate axes. The momentum vector **p** and the electric field vector **E** also behave in this way under reflection. All such vectors are called *polar vectors*.

Now, consider the vector product of two polar vectors, e.g., the angular momentum $\mathbf{L} = \mathbf{r} \times \mathbf{p}$. Each factor changes sign under reflection, but the sign of the product remains unaltered (see Figure 3). The intrinsic angular momentum vector of a particle, i.e., its spin **J**, and the magnetic field vector **H** do not change sign either. All such vectors are said to be *axial*

[⊆] Certain exceptions to this rule, due to the conservation of angular momentum, will be discussed later.

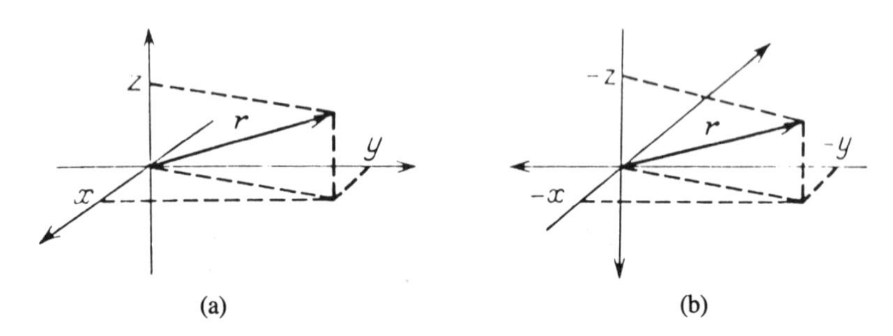

Figure 30. The signs of the components of the position vector are reversed under the transformation from a left-handed to a right-handed coordinate system.

vectors or *pseudovectors*.

The scalar product of two polar vectors (or of two axial vectors) does not change sign under mirror reflection. It is a *scalar*. The scalar product of an axial and polar vector does change sign; it is a pseudoscalar. Scalars are said to be P-even, and pseudoscalars P-odd.

Enough has now been said to make it clear that the emission of longitudinally polarized particles signifies the violation of P symmetry, i.e., P parity is not conserved. For example, the reflection of the β-decay of the neutron in a mirror would be a β-decay with the emission of a right-polarized electron, and this cannot occur in nature.

The reflection of a longitudinally polarized particle in a mirror resembles the reflection of a spinning toy top (Figure 31a) or a screw (Figure 31b). The axis of the top or screw plays the part of a polar vector. Neither the spinning top nor the screw is identical with its mirror image. If necessary, we could make a "mirror-reflected" top; screws with unconventional threads are sometimes made; but right-polarized β-electrons cannot be found in nature: they are invariably left-polarized.

We are not referring here to electrons in general, but specifically to electrons emitted in β-decay. Of course, by passing a beam of electrons through a layer of iron magnetized in the direction of the beam, we can produce either left- or right-polarized electrons simply by reversing the magnetizing field. In this experiment, the polarization of the electrons is "manufactured", like the thread of a screw. However, in β-decay and in other weak processes, the longitudinal polarization and, hence, the mirror asymmetry are *not* introduced from outside: they are intrinsic features of these processes.

The discovery of the nonconservation of parity was completely

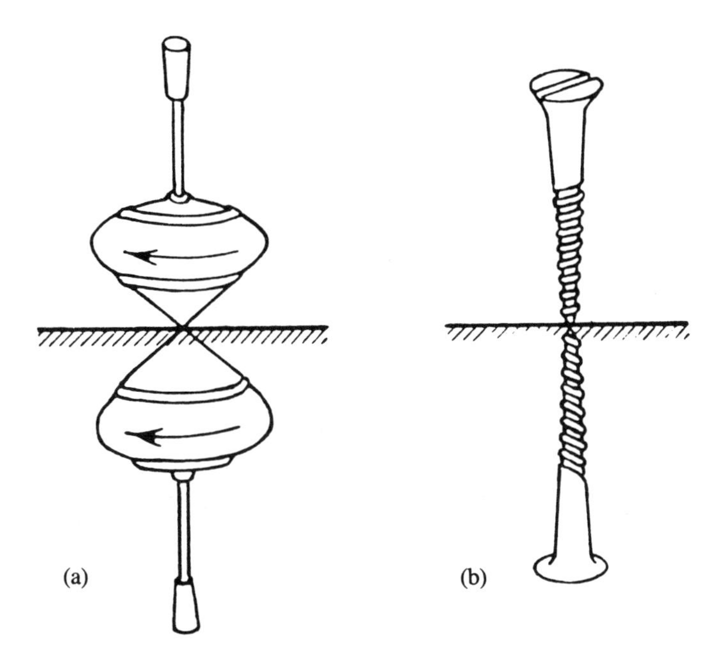

Figure 31. A spinning top (a) and a screw (b) are not identical with their mirror images.

unexpected – like a bolt from the blue. I recall that, in 1954, when I began my postgraduate work under I. Ya. Pomeranchuk, he asked me to calculate the angular distribution of electrons in decays of polarized muons. I also had to devise a method for polarizing the muons.

The calculations of the decay proved to be very cumbersome because a relativistically invariant method of dealing with processes involving polarized particles was unknown at the time. Today, two pages would suffice but, thirty years ago, it took twenty school exercise books to write down all the derivations and I finally obtained a compact (and, as it turned out, correct) expression.

I also tried to invent a method for polarizing muons in a cold magnetized medium (the muon lives for only two microseconds, so that it was necessary to produce its polarization after slowing down in the medium but before it decayed). I failed to think up anything reasonable but, less than two years later, it was found that muons produced in π-meson decays (the main source of muons) were emitted in the state of total longitudinal polarization!

■ The polarizations of muons in the decays

$\pi^+ \to \mu^+\nu_\mu$ and $\pi^- \to \mu^-\tilde{\nu}_\mu$

are -1 and $+1$, respectively (and not $+v/c$ and $-v/c$, as in most other decays). This occurs because the spin of the π meson is zero, and the neutrino and antineutrino are 100% polarized, since they move with the speed of light. It is the fastest competitor that wins a tough race, so the μ^+ is forced to be left-polarized and μ^- right-polarized, in defiance of their usual ways (see Figure 32, where heavy arrows show momenta and light arrows show the spins of the particles).■

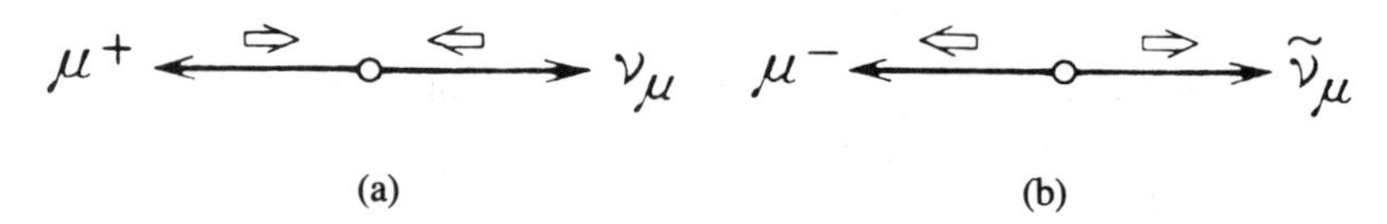

Figure 32. Orientation of the momenta and spins of μ^+ and ν_μ in the decay of the π^+ meson (a) and of μ^- and $\tilde{\nu}_\mu$ in the decay of the π^- meson (b). Spins are shown by the light arrows.

Polarized muons are now widely used in chemistry and solid-state physics. By observing the rotation of muon spins in different media, it is possible to extract information about the properties of these media that cannot be obtained by other methods.

The mirror and charge asymmetry of weak decays is taken into account in the expression for the charged weak current, which includes only left-handed components of leptons and quarks and right-handed components of antileptons and antiquarks. This structure distinguishes the charged weak currents from the electromagnetic and the colored currents, which incorporate left- and right-handed components of particles in a symmetric manner.

■ The electromagnetic and colored currents are four-dimensional vectors. This means that they transform under the Lorentz transformations and under reflections in the same way as the four-dimensional energy-momentum vector $p_\mu = (E/c, \mathbf{p})$ or the four-dimensional space-time vector $x_\mu = (ct, \mathbf{r})$.

A four-dimensional vector V_μ has four components ($\mu = 0, 1, 2, 3,$ where 0 is the time subscript and 1, 2, 3 are the space subscripts; $V_\mu = V_0, \mathbf{V})$. Under the P reflection,

$$V_o \rightarrow +V_o, \ \mathbf{V} \rightarrow -\mathbf{V},$$

so that V_o is a scalar in three-dimensional space and **V** is a three-dimensional polar vector.

In contrast to electromagnetic and colored currents, weak currents are linear combinations of a four-dimensional vector V_μ and a four-dimensional axial vector A_μ.

A four-dimensional axial vector A_μ transforms under the Lorentz transformation in the same way as V_μ, but under P reflections it behaves differently:

$$A_o \rightarrow -A_o, \ \mathbf{A} \rightarrow +\mathbf{A}.$$

It follows that A_o is a pseudoscalar in three-dimensional space and **A** is a three-dimensional axial vector. The weak left-handed charged current is the difference between the curents V_μ and A_μ:

$$L_\mu = \tfrac{1}{2}(V_\mu - A_\mu).$$

The reader should notice that mirror reflection does not transform the components of L_μ into themselves but transforms them into the components of the right-handed current

$$R_\mu = \tfrac{1}{2}(V_\mu + A_\mu).$$

This resembles the way that a right hand looks like a left hand when reflected in a mirror. Clearly, under the P reflection

$$L_o \rightarrow R_o, \ \mathbf{L} \rightarrow -\mathbf{R}. \blacksquare$$

It was hoped, until 1964, that, although the P and C symmetries were completely violated in weak processes, the combined CP symmetry was still valid. The mirror image of the β-decay of the neutron would have then been the β-decay of the antineutron. However, the decay of the so-called long-lived neutral K-meson[c] into two π mesons was discovered in 1964. Symbolically,

$$K_L^o \rightarrow \pi^+\pi^-.$$

---

[c] There are two neutral K mesons with definite lifetimes: the short-lived neutral kaon K_S^o with a lifetime of about 10^{-10} s, and the long-lived K_L^o with the lifetime of about 5×10^{-8} s (K_S^o and K_L^o are two distinct linear combinations of K^o and $\bar{K}^o$, discussed on p. 35). The main decay channels of the K_S^o are the decays into $\pi^+\pi^-$ and $\pi^o\pi^o$. The main decay channels of the K_L^o are the decays into $\pi^o\pi^o\pi^o$, $\pi^+\pi^-\pi^o$, $\pi^\pm e^\mp \nu$, and $\pi^\pm \mu^\mp \nu$.

It can be shown that, if the CP symmetry holds, this decay must be forbidden. Another CP-forbidden decay was subsequently discovered, namely,

$$K_L^o \to \pi^o\pi^o.$$

A slight charge asymmetry in the decays

$$K_L^o \to \pi^\pm e^\mp \nu \text{ and } K_L^o \to \pi^\pm \mu^\mp \nu$$

was also found. In the latter case, the probability of decays resulting in the emission of positive leptons (e^+, μ^+) was about 0.3% higher than that of decays into negative leptons (e^-, μ^-).

The decays of K_L^o mesons further revealed that T symmetry was also violated. So far the violation of the CP and T symmetries has been observed only in decays of K_L^o mesons. The nature of this symmetry breaking and its mechanism are still unclear, although a number of theoretical models have been suggested to explain the phenomenon.

The only discrete symmetry that remains unbroken is CPT symmetry, i.e., the symmetry under the product of all three transformations, C, P, and T. This symmetry is so profoundly incorporated in the foundations of modern field theory that most theoretical physicists adamantly believe it is unassailable. We note that the CPT symmetry alone is sufficient to ensure that a particle and its antiparticle have equal masses as well as equal lifetimes.

Neutral currents

So far, whenever we referred to weak processes, we invariably meant interactions between charged currents. A new type of weak process was discovered in 1973, namely, the interaction between *neutral currents*.

Neutral currents are said to be currents that do not change the charges of leptons and quarks, for example, $\bar{\nu}_\mu \nu_\mu$, $\bar{e}e$, $\bar{u}u$, $\bar{d}d$, and so on. Moreover, the experimentally observed neutral currents are truly neutral: they leave unchanged not only the charges but also all other quantum numbers of the particles. They are said to be diagonal, i.e., they transform a lepton or quark into itself. Nature has no nondiagonal neutral currents like $\bar{d}s$ or $\bar{e}\mu$, or $\bar{u}c$.

Inspection of Figures 28 and 29 will readily show that neutral currents are diagonal. As required by the conservation of electric charge, neutral currents cannot transform particles from different triples into one another. Since the unit vectors of particles within each triple are mutually orthogonal, each of the twelve particles can transform only into itself (when color is taken into account, this number is not 12 but 24).

It took three-quarters of a century to discover neutral currents after the discovery of the β-decay, precisely because these currents did not change the flavors of the particles. They cannot convert fundamental fermions into one another and, therefore, cannot be seen in particle decays in which the flavors of leptons or quarks are changed. If a decay is flavor-conserving, the weak interaction of neutral currents cannot compete with the electromagnetic interaction because the latter is much stronger.

Neutral currents were first found in neutrino experiments on large accelerators in which a high-energy neutrino beam passed through a target and produced reactions of the form

$$\nu_\mu + \mathrm{p} \rightarrow \nu_\mu + \mathrm{p} + \pi^+ + \pi^-.$$

Such reactions are the result of the shake-up of a nucleon when a neutrino interacts with one of the quarks: $(\bar{\nu}_\mu \nu_\mu)$ $(\mathrm{u}^-\mathrm{u})$ or $(\bar{\nu}_\mu \nu_\mu)$ $(\bar{\mathrm{d}}\mathrm{d})$. The probabilities of these reactions were found to be nearly the same as the probabilities of similar reactions produced by charged currents, for example, the reaction

$$\nu_\mu + \mathrm{n} \rightarrow \mu^- + \mathrm{p} + \pi^+ + \pi^-.$$

Neutral currents (like charged currents) violate mirror symmetry. It was as a result of a search for the breaking of this symmetry that the neutral electron current $\bar{\mathrm{e}}\mathrm{e}$ was discovered in 1978. This was a much more complicated task than the detection of the neutral neutrino current because the effects of the weak electron current had to be separated from the much stronger electromagnetic interaction of the electron.

First, an extremely small rotation of the plane of polarization of laser light was discovered in atomic bismuth vapor. This rotation is obviously P-odd because neither the left nor right sense of rotation would be in any way distinguished for the completely P-symmetric bismuth atoms, and the plane of polarization of the laser light, like Buridan's ass, would stay fixed.

The rotation of the plane of polarization of light in solutions of mirror-asymmetric molecules is a well-known phenomenon. For example, this rotation is observed when a beam of polarized light is passed through a glass of sweetened tea. Of course, this elementary school experiment does not reveal any parity violation at the level of fundamental interactions. It merely indicates that sugar molecules are asymmetric.

Mirror twins of ordinary sugar molecules can be synthesized in the laboratory, but the sugar beet or cane refuses to produce it. Here, we encounter a special case of a striking phenomenon: all living organisms exhibit mirror asymmetry. This is a fascinating topic for discussion, but it would take us too far off course.

Let us return to atomic bismuth. The point here is that we are dealing with atoms and not molecules. If P parity were conserved, bismuth atoms would be mirror symmetric. A slight "twist" is caused by the P-odd weak interaction between atomic electrons and the nucleus of the bismuth atom (to the accuracy within which CP symmetry holds, the mirror twin of the bismuth atom is the bismuth antiatom; we have no means for creating such heavy antiatoms and, besides, what use would they be if we could?).

Soon after the results of the bismuth experiment became known, another experiment produced similar results, this time in an electron accelerator. The experimenters were observing the interaction between a beam of longitudinally-polarized electrons and a deuterium target. Left-polarized electrons were found to be scattered by deuterium about one-hundredth of one percent more effectively than right-polarized electrons. It is not difficult to understand that this effect provides further evidence for the breaking of mirror symmetry in the interaction between electrons and quarks. Calculations show that the observed effect is in quantitative agreement with the hypothesis that the asymmetry is caused by the weak interaction between neutral currents.

Neutral current effects were also observed in 1982, in the reactions $e^+e^- \rightarrow \mu^+\mu^-$ and $e^+e^- \rightarrow \tau^+\tau^-$ in colliding electron and positron beams.

In agreement with theory, all leptons and quarks have the corresponding neutral currents. All neutral currents of neutrinos are as left-handed as charged currents. The neutral currents of all other particles have both left-handed and right-handed components, L_μ and R_μ.

Predicted W and Z bosons

The theory of interaction between vector bosons and neutral and charged currents, now known as the standard theory of the electroweak interaction (this term will be explained below), was developed in the 1960s. In 1979, three theorists, S. Glashow, A. Salam, and S. Weinberg, won the most prestigious international science award, the Nobel Prize, for this outstanding achievement. The prize was awarded four years before the W and Z bosons were actually discovered, i.e., before the validity of the theory was finally confirmed. The point is that the discovery of neutral currents in 1973, and the experimental data on their structure accumulated in subsequent years, provided a very conclusive verification of the validity of the theory.

According to the standard theory, all interactions between charged currents proceed via the exchange of W bosons, and all interactions between neutral currents proceed via the exchange of Z bosons. Like the photon and the gluons, they are vector bosons, i.e., spin-1 particles.

The coupling constant g_W of the interaction between W bosons and charged currents plays a role similar to that of the electric charge e in electrodynamics. By analogy with the fine structure constant $\alpha = e^2/\hbar c$, we can introduce the quantity $\alpha_W = g_W^2/\hbar c$. The ratio α/α_W is usually written in the form

$$\alpha/\alpha_W = \sin^2\theta_W,$$

where θ_W is the so-called weak, or Weinberg, angle. Experimental studies of neutral weak currents have shown (see below) that $\sin^2\theta_W = 0.21–0.23$ and $\alpha_W \simeq 1/30$. Note that the weak charge g_W is greater than the electric charge e. The fact that, as a rule, the weak interaction is much weaker than the electromagnetic interaction is explained not by the smallness of the weak charge but by the fact that W bosons are very heavy.

Once α_W is known, we can express the W-boson mass in terms of α_W and the Fermi constant G_F (we recall that $G_F \simeq 1.2 \cdot 10^{-5}\ \mathrm{GeV}^{-2}\hbar^3c^3$). For example, consider the decay of the muon (Figure 33). The momentum q of the virtual W boson in this process (divided by the speed of light c) is not greater than the muon mass and is therefore smaller by many orders of magnitude than the mass of the W boson. The propagator describing the motion of a virtual vector boson has the same form as the propagator of a "massive photon" (it was discussed earlier in the book), i.e.,

$$\frac{1}{m_W^2 - q^2}.$$

When $|q^2| << m_W^2$, we can neglect q^2 in comparison with m_W^2, so that

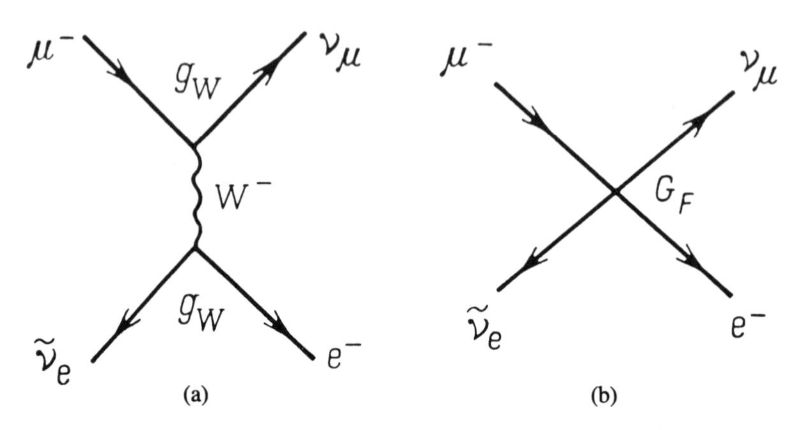

Figure 33. Feynman diagram describing the muon decay proceeding via the creation and decay of a virtual W boson (a) and the same diagram in the approximation of four-fermion interaction (b). g_W is the weak charge and G_F is the Fermi constant.

$$\frac{1}{m_W^2 - q^2} \simeq \frac{1}{m_W^2}.$$

This approximation holds for the virtual W boson in the muon decay to within $10^{-3}\%$. This is the accuracy with which we can replace the "true" diagram of Figure 33a with the approximate diagram of Figure 33b, which shows a four-fermion interaction.

The following comparison may be helpful here. A virtual particle is like a spring. For example, consider a virtual photon. It is described by the propagator $-1/q^2$. When q^2 is large, the "photon spring" is short, and when q^2 is small, the spring is long. Indeed, as prescribed by the uncertainty relation, the length l is given by $l \approx \hbar/q$. The "spring" describing a heavy W boson is always short ($l \lesssim h/m_W c \simeq 2 \cdot 10^{-16}$ cm) and stiff. As long as the momentum transferred to the spring is small, the spring remains practically uncompressed and acts like a rigid body. This situation represents the limiting case obtained by neglecting q^2 in the propagator. The above ideas are consistent with the uncertainty relation and with the formula for the potential due to the exchange of a heavy virtual photon (see the section entitled *Virtual particles*).

Let us now evaluate the mass of the W boson. The diagrams in Figure 33 can be used to show that

$$2\sqrt{2}G_F = \frac{(g_W/\sqrt{2})^2}{m_W^2}.$$

The functional relation between G_F, g_W, and m_W is obvious from the comparison of the diagrams of Figures 33a and 33b. As for the numerical coefficients, you will easily derive them if and when you become a particle physicist.

This relation can also be rewritten in the form

$$m_W = \left(\frac{\sqrt{2}G_F}{\pi \alpha_W \hbar^3 c^3} \right)^{-\frac{1}{2}} \simeq \frac{37.3 \text{ GeV}}{\sin \theta_W}.$$

For $\sin^2\theta_W = 0.21$, we then have $m_W = 81.4$ GeV.

Thus, once $\sin^2\theta_W$ became known from experiment, this was sufficient for predicting the mass of the W boson!

We have already mentioned that $\sin^2\theta_W$ was found by studying neutral currents. The point is that, according to the standard theory, the angle θ_W characterizes the interaction not only of W bosons but also of Z bosons. In the latter case, θ_W appears in the expressions for the following three physical quantities:

(1) the mass of the Z boson

$$m_Z = m_W/\cos\theta_W$$

(this gives $m_Z = 91.6$ GeV for $m_W = 81.4$ GeV);

(2) the coupling constant between the Z boson and the neutral current

$g_Z = g_W/\cos\theta_W$; and

(3) the neutral current itself, which has the following form for a particle with electric charge Q:

$$\pm \tfrac{1}{2}L_\mu - (L_\mu + R_\mu)\, Q\sin^2\theta_W,$$

where L_μ and R_μ are the left- and right-handed currents. The plus sign in the first term corresponds to the "upper" leptons and quarks (ν_e, ν_μ, ν_τ, u, c, t), and the minus sign corresponds to the "lower" particles (e^-, μ^-, τ^-, d, s, b).

It follows directly from the expressions for m_Z and g_Z that $g_Z^2/m_Z^2 = g_W^2/m_W^2$ and that the four-fermion interaction constant for neutral currents is the same as for charged currents. The entire dependence of the probabilities of processes caused by neutral currents on $\sin^2\theta_W$ is therefore determined by the above expression for the neutral current. It shows that the degree of longitudinal polarization, i.e., the ratio of the right- and left-handed components, is different for particles with different charge and is a function of $\sin^2\theta_W$. Experiment shows that the neutral currents of neutrinos ($Q_\nu = 0$), electrons and muons ($Q_e = Q_\mu = -1$), and quarks ($Q_u = +\frac{2}{3}$, $Q_d = -\frac{1}{3}$) fit the above expression for the neutral current when $\sin^2\theta_W \simeq 0.21$.

As an example, Figure 34 shows a diagram describing the scattering of a muon neutrino by an electron. The cross section for this process can be used to determine $\sin^2\theta_W$. We have already noted that the standard theory was thoroughly tested by the experimental studies of other weak processes as well (some are shown in Figure 35).

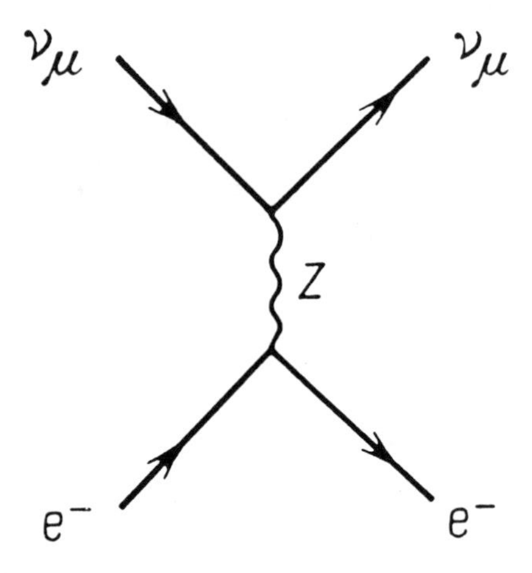

Figure 34. Feynman diagram describing the scattering of a muon neutrino by an electron via the exchange of a virtual Z boson.

The standard theory of the electroweak interaction is based on the so-called electroweak symmetry, which involves four massless vector bosons: two charged and two neutral. Unlike color symmetry, electroweak symmetry is broken in nature. The result of this symmetry breaking is that there is only one massless vector boson, i.e., the photon. The other three, W^+, W^-, and Z^o, have nonzero masses.

The word "electroweak" is used to describe both weak and electromagnetic interactions (we recall that $\sin\theta_W = e/g_W$). The existing electroweak theory has not, alas, achieved the final unification of the electromagnetic and weak interactions because $\sin^2\theta_W$ is taken from the experiment rather than predicted by the theory itself from some very general principles.

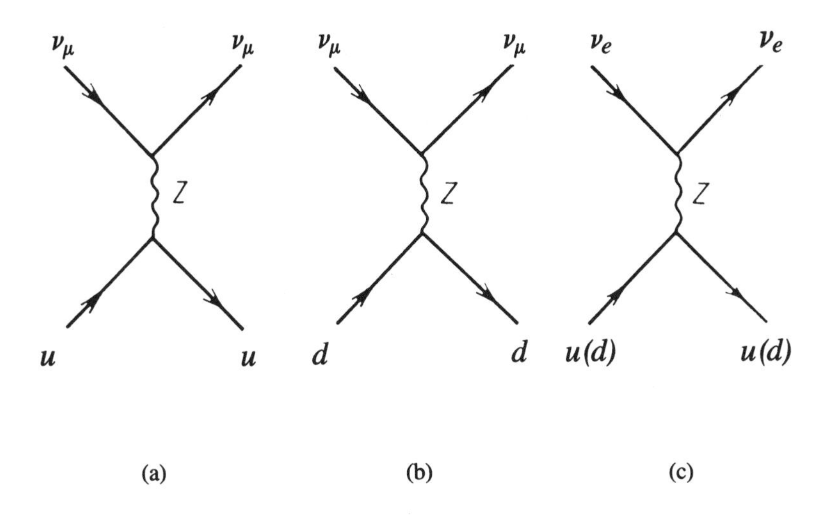

Figure 35. Scattering of a muon neutrino [(a) and (b)] and of an electron neutrino (c) by u and d quarks.

The discovery of W and Z bosons

The energy required to create heavy particles such as the W and Z bosons is so high that it could not be attained in particle collisions in accelerators that had been available until the early 1980s. Actually, two laboratories, the Fermi National Accelerator Laboratory (Fermilab) and the European Organization for Nuclear Research (CERN) did have at their disposal beams of particles with energies exceeding the mass of the intermediate bosons by a factor of 5 or 6, but these beams were made to collide with fixed (stationary) targets, so that the energy available for the production of new particles was only a small fraction of the energy E of the beam particles. This fraction is $\sqrt{}(2mc^2/E)$, where m is the nucleon mass. Colliding beams were obviously necessary if intermediate bosons were to be produced.

In 1976, D. Cline, P. McIntyre, and C. Rubbia suggested that proton-antiproton colliders based on the Fermilab and CERN proton accelerators should be built. A decision to implement this plan was made at CERN in

1978, and a collider in which 270-GeV protons collided with antiprotons of the same energy became operational in 1981. At first, the intensity of the colliding beams was too low for the intermediate bosons to be detected (collisions were too rare). However, by the fall of 1982, the necessary threshold was exceeded and, in January 1983, Carlo Rubbia, the leader of the experimental group working on the so-called UA1 detector, announced the observation of the first W bosons.

How are intermediate bosons created? A fast proton can be thought of as a beam of its component particles, i.e., quarks and gluons. Roughly half the momentum of a fast proton is carried by quarks and the other half by gluons. Similarly, a fast antiproton is a beam of antiquarks and gluons. When a proton and antiproton collide, one of the quarks (q) and one of the antiquarks ($\tilde{q}$) may transform into an intermediate boson (see Figure 36). Four such quark-antiquark reactions are possible. They are shown in

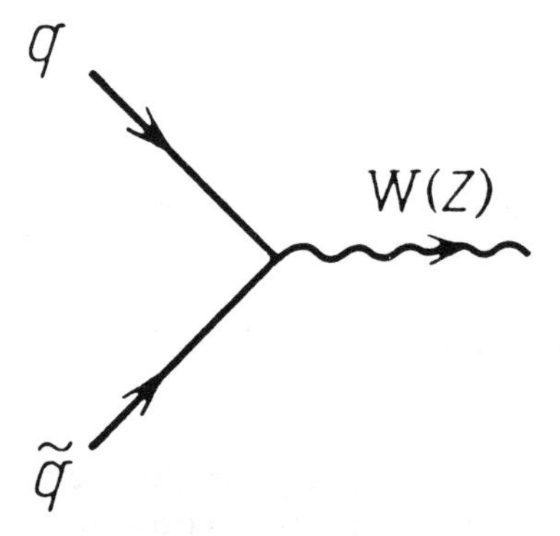

Figure 36. Creation of an intermediate boson (W or Z) in a collision between a quark and antiquark.

Figure 37. The other quarks, antiquarks, and gluons take no active part in the creation of intermediate bosons: they are merely passive spectators. However, once they have lost their companions, the spectators experience a "profound shock" and fragment into hadrons, which emerge in the form of two hadron jets, one of which follows the trajectory of the proton and the other that of the antiproton. This collision is illustrated in Figure 38, where the straight lines represent quarks and antiquarks and the wave line represents the W (or Z) boson; gluons are not shown.

Note that the line representing the intermediate boson in Figure 38 is very short, which is meant to indicate that, according to the standard

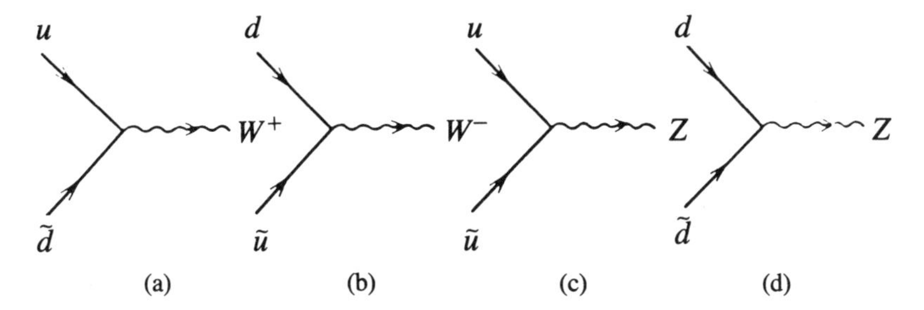

Figure 37. Production of a W^+ boson (a), W^- boson (b), and Z boson (c, d) in quark-antiquark collisions.

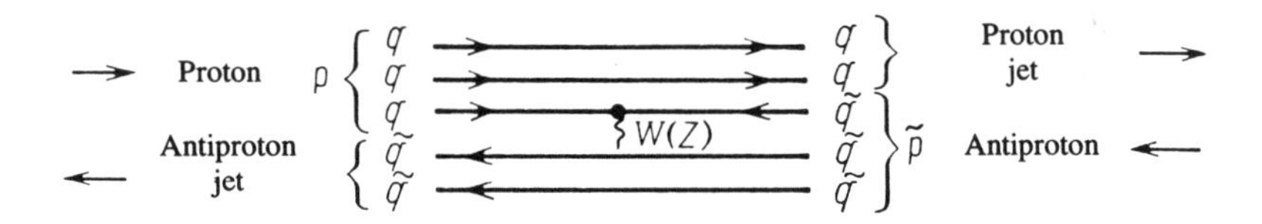

Figure 38. Production of an intermediate boson (W or Z) in a proton-antiproton collision. The spectator quarks and antiquarks form two hadronic jets.

theory, the intermediate boson lives for only $3 \cdot 10^{-25}$ s and does not even get clear of the crowd of quarks and antiquarks that witnessed its birth. How, then, is it possible to ascertain that an intermediate boson was created at all? The answer is that a decaying intermediate boson sometimes sends out an exceptionally clear signal.

In most cases, an intermediate boson decays into a quark and antiquark, which then transform into two hadron jets. Unfortunately, it is very difficult to differentiate between the hadrons in these jets and hadrons created in a collision due to the strong interaction between the quarks in the proton and antiproton. In contrast, the decays of the W and Z bosons into leptons are not swamped by any background. The decays

$$W^+ \rightarrow e^+\nu_e,\ W^- \rightarrow e^-\tilde{\nu}_e,\ Z^o \rightarrow e^+e^-,$$

in which electrons and positrons emerge at a large angle to the colliding p and $\tilde{p}$ beams, are examples of this.

The detectors in which W and Z bosons were observed are quite unique. They combine giant size with extreme precision. Figures 39, 40, and 41 show schematically the UA1 detector. Figure 39 is a lateral view along the

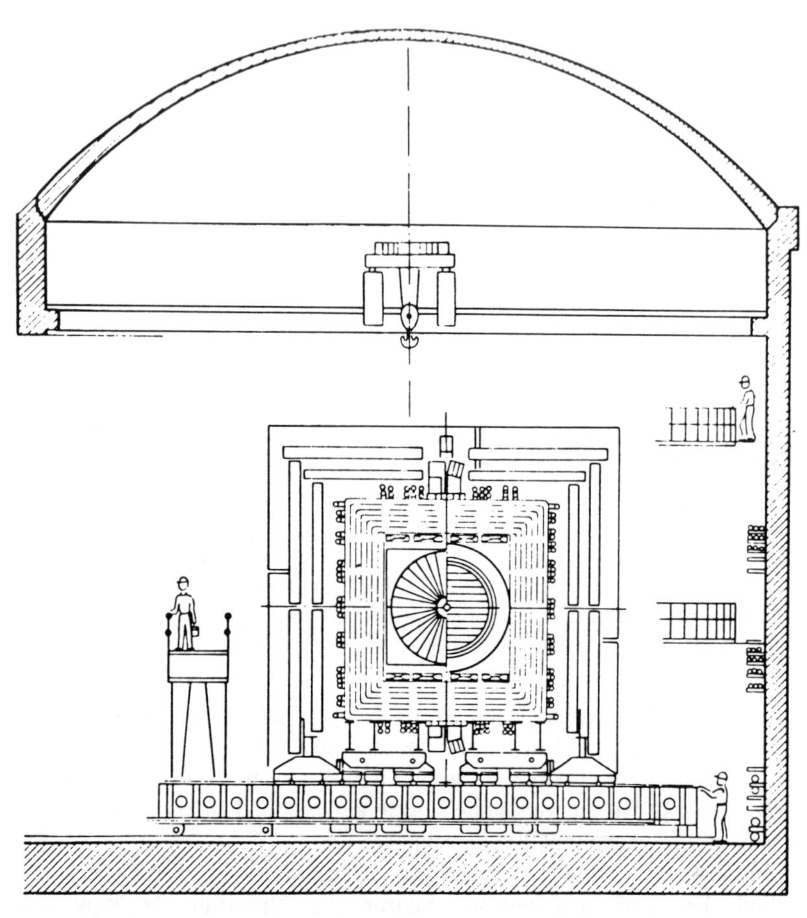

Figure 39. Schematic lateral view of the UA1 detector (the proton beam is perpendicular to the page).

proton beam, whilst Figure 40 shows the front view, with protons entering the detector from the left and antiprotons from the right. Figure 41 shows the detector partially disassembled, with the cylindrical central detector in a 0.7 Tesla magnetic field, generated by an 8000-ton magnet. The internal volume of the magnet is 80 m^3, of which 25 m^3 is occupied by the central detector (5.8 m long and 2.3 m in diameter).

The central detector is the so-called *drift chamber*. It is filled with a gaseous mixture of argon and ethane and contains 6000 very thin wires

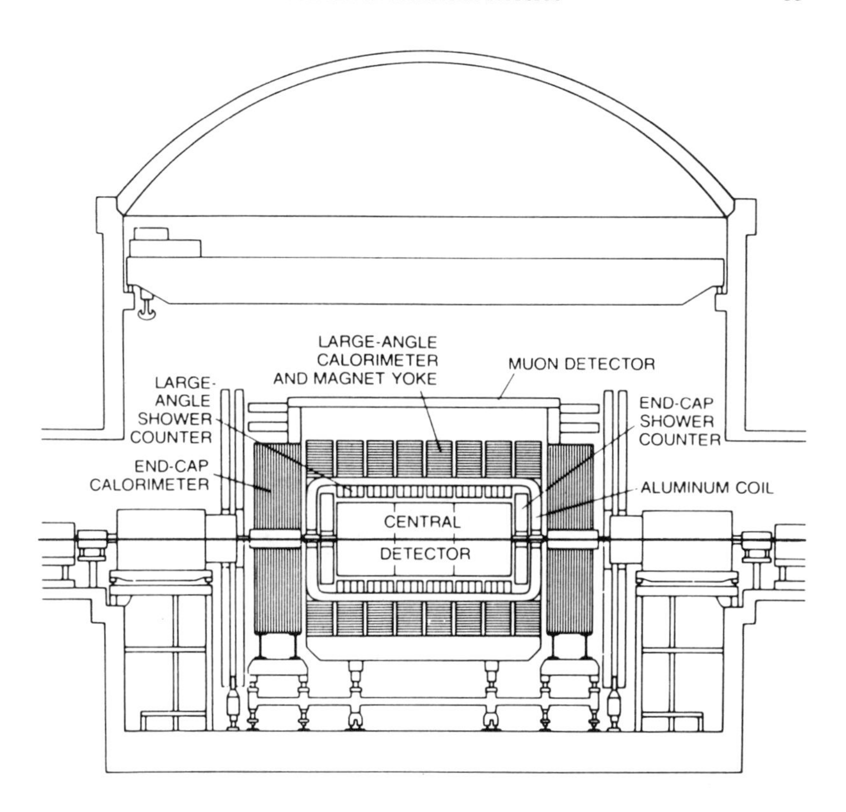

Figure 40. Schematic front view of the UA1 detector.

that record electrons knocked out of the gas atoms by ionizing particles. These wires form planes separated by 18 cm. A further 17 000 wires produce the electric field in which the ejected electrons drift with a velocity of 5.3 cm/μs. The maximum electron drift time is therefore 3.6 μs, i.e., less than the interval (3.8 μs) between successive collisions of the proton and antiproton bunches in the vacuum tube of the collider, which runs along the axis of the central detector. Signals from the detector wires are recorded by fast electronics. An on-line computer reconstructs the particle trajectories and calculates their momenta.

The central detector is surrounded by an electromagnetic calorimeter that absorbs the electrons, positrons and photons and measures their

Figure 41. Axonometric view of the UA1 detector moved away from the beam, with two parts of the magnet separated.

energy. The calorimeter consists of alternate layers of scintillator and lead. The very word "calorimeter" indicates that the system measures the total energy of the particles.

High-energy hadrons pass through the electromagnetic calorimeter without depositing appreciable amounts of energy in it. Their energy is measured by the hadron calorimeter, which is actually the yoke of the magnet with layers of a scintillator imbedded within it.

Finally, the whole complex is surrounded by 50 drift-chamber plates (4m by 6m each), which record muons.

Two vertical silos, each 20 m in diameter, were constructed for the UA1 experiment. Collider beams pass horizontally through one of them, whilst the other serves as a "garage" for the detector complex when the collider is turned off. The whole UA1 detector weighs 2000 tons and is transported from one silo to another on rails.

The second intermediate boson detector, the UA2, does not have a large magnet. It is smaller and lighter than the UA1 and moves on an air cushion in the experimental hall, 50 m below ground level (the letters UA stand for Underground Area). All in all, six experimental groups, UA1 through UA6, worked on the CERN $p\bar{p}$ collider in 1983–1984.

The scale of these experiments is illustrated by the fact that, when the discovery was reported by the UA1 group, the paper had 137 authors!

The main difficulty in the search for intermediate bosons is that they are produced with extremely low probability. Thus, the number of events of the form

$$p\bar{p} \rightarrow W + \text{hadrons}, \quad W \rightarrow e^{\pm}\nu$$

and

$$p\bar{p} \rightarrow Z + \text{hadrons}, \quad Z \rightarrow e^{+}e^{-}$$

is, respectively, about 10^{-8} and about 10^{-9} of the total number of events occurring when protons collide with antiprotons. The intermediate boson needle must thus be found in the haystack of billions of strong interaction events, each of which creates dozens of particles. These large numbers invite another metaphor: it is like looking for one person among the five billion population of the world.

Of the billion $p\bar{p}$ collisions that occurred in the UA1 detector during a 30-day experimental run in November–December 1982, the fast electronics discarded 999 million as "not promising" and sent to the computer only one million of the events. These were the events involving particles with high transverse energies, i.e., high-energy particles emerging at large angles. The analyzing electronics then selected about 1000 events with transverse energies greater than 15 GeV, and a thorough analysis of these thousand finally produced six candidates for the status of W bosons.

We have already mentioned above that the hallmark of the decay of a W boson is an electron with high transverse energy. When a W boson with a mass of about 80 GeV decays, an electron and a neutrino share the available energy almost equally, so that each acquires kinetic energy of about 40 GeV. When an electron with this energy emerges at $90° \pm 30°$ to the beam axis, its transverse energy exceeds 30 GeV. This criterion was satisfied by four of the six possible events.

Another characteristic feature of the $W \rightarrow e\nu_e$ decay is the neutrino. Because of conservation of momentum, the transverse momentum of the neutrino must be equal in magnitude and opposite in direction to the transverse momentum of the electron (the small transverse momentum of the decaying W boson can be ignored). The penetrating power of the neutrino is very high (the range of a 40-GeV neutrino in iron exceeds one million kilometers), so that it escapes from the UA1 detector without leaving a track behind it. This absence of tracks is the main evidence for the presence of the neutrino. Indeed, all other particles either stop in the system (electrons, photons, hadrons) or at least leave a track behind (muons). The evidence for a neutrino emitted with high transverse momentum is thus the apparent violation of the conservation of transverse momentum, precisely by the amount corresponding to the transverse momentum of the neutrino. The transverse momentum in four of the above W events was indeed deficient by 30–40 GeV!

Soon after the announcement from the UA1 group, the UA2 group also reported the observation of their first W-boson events.

After the next collider run in April–June 1983, the total statistics of W-boson events grew to almost one hundred, and about ten electron-positron decays of the Z boson were identified. The measured masses of the W and Z bosons were just over 80 and 90 GeV, respectively. The experiment also yielded the boson production cross sections and the angular distributions of decay products. These were found to be in agreement with the predictions of the standard theory.

The discovery of the W and Z bosons is a triumph for the theory that would be impossible without the extraordinary skill of the accelerator engineers and experimental physicists. In 1981, the CERN Publication Group published a booklet entitled "The CERN Antiproton Project", which began the history of the project with the following lines:

1967 The technique of electron cooling is invented at the Institute of Nuclear physics, Novosibirsk, USSR, with a view to achieving intense antiproton beams for proton-antiproton colliding beam physics at an energy of 25 GeV.

1968 The technique of stochastic cooling is invented at the European Organization for Nuclear Research (CERN) with the initial aim of improving beam quality . . .

The "father" of electron cooling was the Soviet physicist, Academician G.M. Budker (1918–1977). Stochastic cooling was invented by the Dutch physicist, Simon van der Meer. We shall not describe here the principles of the two methods. Suffice it to say that the $p\bar{p}$ collider cannot be made to work without cooling the antiproton beam. It was thus no surprise that the 1984 Nobel Prize for physics was awarded, for the discovery of intermediate bosons, to the experimentalist, Carlo Rubbia, and the accelerator physicist, Simon van der Meer. The reader will recall that theorists were given the Nobel Prize for intermediate bosons as early as 1979. Such a large number of scientists and engineers with different specialities had never before joined forces so harmoniously in the pursuit of a fundamental discovery in physics.

It will be instructive to compare the discovery of the Z boson with the discovery of its closest relative, the photon, at the beginning of this century. We live in an ocean of light: photons literally "hit the eye". The main difficulty in discovering photons lay not in creating them but in having to find the individual particles, i.e., the quanta. In the case of Z bosons, there were a mere dozen to start with, and the main difficulty was with the creation of the particles. This difference stems from the fact that the photon is massless whilst the Z boson is very heavy. If the Z boson were equally massless, it would be easier to observe than the photon. Indeed, the charge g_Z is larger than the electric charge, e.

The W and Z bosons are still under investigation on the CERN $p\bar{p}$ collider. In 1986, it will be joined by the Tevatron, the Fermilab $p\bar{p}$ collider, in which the energies of protons and antiprotons will reach 1 TeV = 1000 GeV.

The next to come on-line will be the "Z-boson factories" – the SLC (1987) and LEP (1988) electron colliders. The first of these is being built as an extension of the Stanford linear accelerator. It will accelerate electrons and positrons synchronously and make them collide head-on at the head of a "racket" (Figure 42). The second collider is circular, with a ring circumference of 27 km. It is being built at CERN and is expected to produce 20 to 30 Z bosons per minute. The energy of electrons and positrons (about 46 GeV per particle) will be adjusted in both colliders so that the colliding e^+ and e^- will produce a Z boson by a resonance mechanism.

The gigantic scale of the LEP is well matched by the sizes of the four detectors that will be installed there. We shall mentioned only one of them, the L3 detector (incidentally, a group of Moscow physicists has taken part in its design and construction, together with their colleagues from China, France, East and West Germany, Hungary, India, Italy, The Netherlands, Spain, Sweden, Switzerland, and the USA). The L3 detector will be equipped with a 6000-ton electromagnet, 16 × 16 × 16 m in size, with an internal volume of 12 × 12 × 12 m^3. This space will house a 300-ton hadronic calorimeter, made of uranium and containing a 10-ton electro-

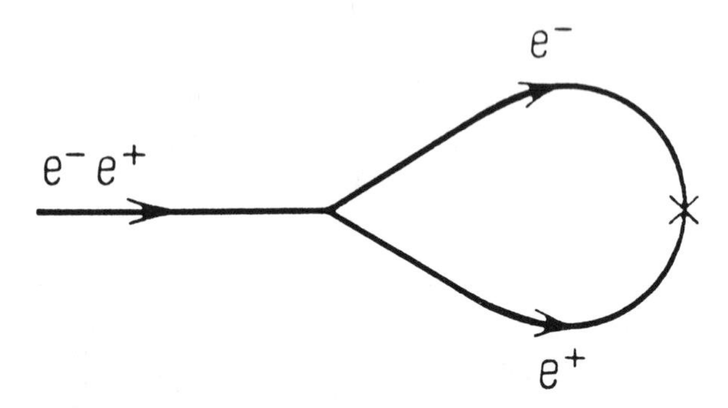

Figure 42. Schematic of particle trajectories in the Stanford linear collider.

magnetic calorimeter composed of 12 000 scintillators in the form of single crystals of BGO ($Bi_4Ge_3O_{12}$).

Post-Z collider physics

Following the discovery of the Z boson, the number one problem in high-energy physics is the search for scalar bosons, i.e., the fundamental spin-0 particles. These particles are called the *Higgs bosons*, after the British theorist, Peter Higgs.

Higgs bosons, often called simply *higgses*, play a very important role (actually, several roles) in current theoretical physics. First, they give leptons, quarks, and intermediate bosons their masses. Second, they are responsible for the "rotation" of the upper quarks relative to the lower quarks (cf. Figure 29) and for the existence of nine (instead of three) charged quark currents. Third, they seem to cause the violation of the CP and, possibly, P symmetries in weak processes. Unfortunately, it is extremely difficult to explain simply how the Higgs bosons manage to do all this. A further book would be needed (and many such books will undoubtedly be written after these particles have been discovered).

The theory of Higgs bosons has not yet reached the status achieved fifteen years ago by the theory of intermediate W and Z bosons. It still contains too many free (arbitrarily chosen) parameters, and its most significant shortcoming is that it gives no clear predictions about the mass (or masses) of these particles. The lower bound is a few GeV. The upper bound is about 1000 GeV.

If the Higgs bosons are light, they may be discovered in existing colliders or the LEP. If they are heavy, we shall have to wait until the 1990s, when the supercolliders become operational, namely, the UNK – an accelerator-storage facility at Protvino (near Serpukhov, USSR), producing colliding protons at 3 TeV, and the SSC (Superconducting Supercollider) in the USA. The parameters of the latter project have not yet been finalized, but the proton energy is expected to be about 20 TeV. In addition, there are plans to build, in the 1990s, a 10-TeV proton-antiproton collider with superconducting magnets in the LEP tunnel.

Preliminary studies are in progress now for a completely new type of collider with linear colliding electron-positron beams of 500 GeV each, to be built at Novosibirsk (this is the VLEPP machine, VLEPP being the Russian abbreviation for colliding linear electron-positron beams).

Some physicists hope that these machines may bring about the discovery of the so-called *preons*. These are hypothetical particles that, according to some theoretical models, are the constituents of both leptons and quarks. It may even be the case that preons are the building blocks of not only leptons and quarks but also of some, or even all, vector bosons (W, Z, gluons, photons, and Higgs bosons).

Preons have been suggested by the abundance of fermions and bosons, which are regarded as fundamental at present, and by the great diversity of their properties. However, speculations about preons have not produced a convincing theory. A fundamental principle is probably still lacking.

High-energy colliders will also be needed to verify one further invention of theorists, namely, *supersymmetric particles*. Supersymmetry is the symmetry between fermions and bosons. In the simplest version of supersymmetry, each of the presently known particles has a "superpartner" with spin differing by $\frac{1}{2}$, i.e., there is the *photino* (a spin-$\frac{1}{2}$ particle) for the photon, *leptinos* and *quarkinos* (spin-0 particles) for leptons and quarks (more often, these scalar particles are referred to as *sleptons* and *squarks*), the *higgsino* (a spin-$\frac{1}{2}$ particle) for the Higgs boson, and so on. Clearly, if supersymmetry is realized in nature at all, it is very badly broken. These superpartners (the "inos") are expected to be quite heavy, which would explain why they have not been discovered in existing accelerators. Unfortunately, there are no clear predictions about their masses, but there are arguments indicating that "inos" cannot be much heavier than 100 GeV.

In addition to the hazardous pursuit of the strange products of theorists' imagination (hazardous because success is not guaranteed), physicists working with colliders have a sure way of "delivering the goods", namely, through the quantitative study of strong interactions between gluons and quarks. Experimental studies of hadron jets, which have already been mentioned several times on the preceding page, are of particular interest.

Electron-positron colliders produce mostly the so-called *quark jets* of hadrons, which originate from the creation of a quark-antiquark pair (Figure 43a_1). As they fly away from each other, the high-energy quark and antiquark generate a large number of quark-antiquark pairs (Figures 43a_2, a_3, . . .), so that, by the time they have reached large distances, the colored particles have become the components of colorless complexes that subsequently evolve into hadrons. This results in very narrow beams of hadrons (the so-called jets), which are emitted in the directions of motion of the original quarks (Figure 43b). The angular distribution of the jet axes is the same as the angular distribution of the original parent quark-antiquark pair.

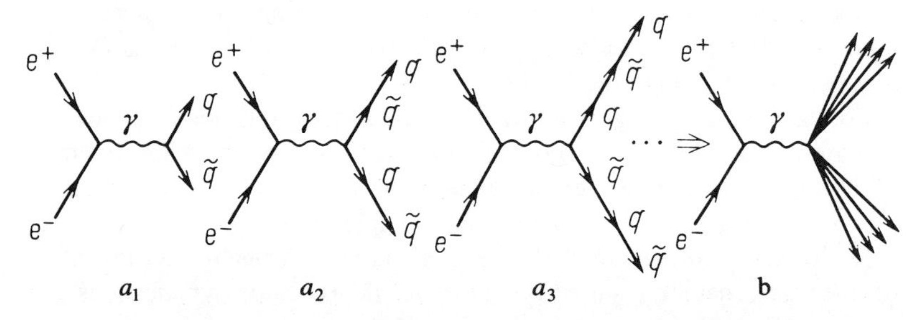

Figure 43. Formation of two hadronic jets by fast quark and antiquark produced in a high-energy electron-positron collision.

Occasionally, whilst the original parent quarks are still at a very short distance, one of them emits a high-energy gluon at a large angle (Figure 44a), which then initiates another jet of hadrons called the *gluon jet* (Figure 44b). (As was mentioned above, a hadron jet initiated by a quark is usually referred to as the quark jet). The differences between quark and gluon jets of hadrons are of great interest.

Presumably, gluon jets are produced in proton-antiproton and proton-proton colliders even more often than quark jets. Two quark jets are produced in large-angle quark-quark and quark-antiquark scattering. Two

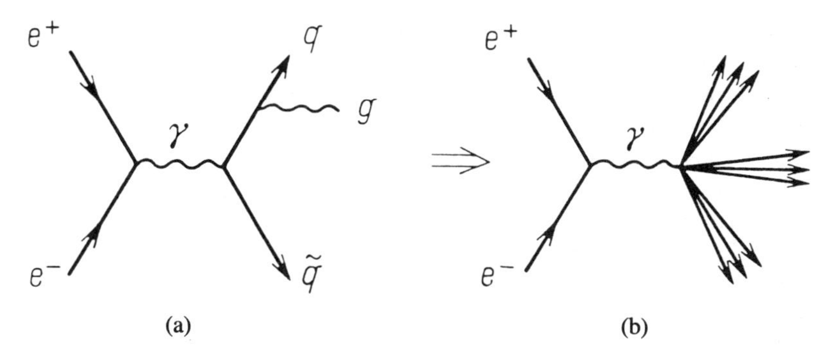

Figure 44. Emission of a gluon by a quark (a) and the resulting formation of three hadronic jets due to a quark, antiquark, and gluon, respectively (b).

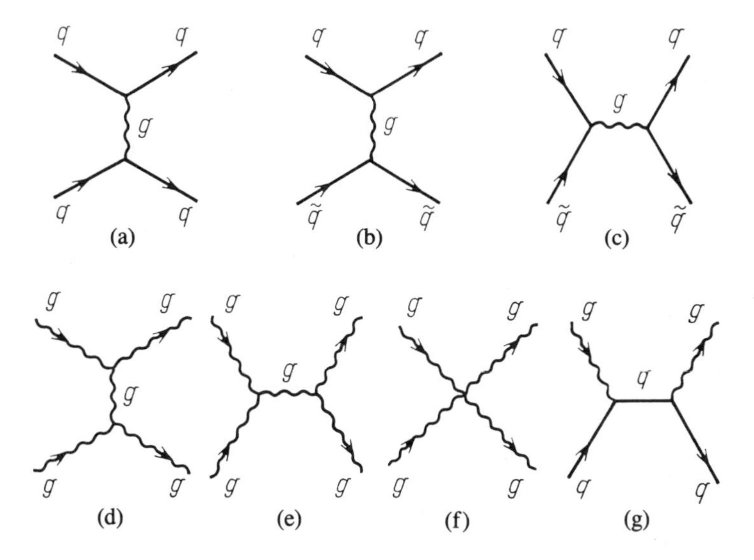

Figure 45. Scattering of a quark by a quark (a), a quark by an antiquark (b, c), a gluon by a gluon (d, e, f), and a gluon by a quark (g).

gluon jets are created in gluon-gluon scattering. A quark and a qluon jet are produced simultaneously in quark-gluon scattering (Figure 45).

We have already noted that gluons carry about half the momentum of a proton. Gluon jets are born in high-energy collisions more frequently because the probability of a gluon-gluon collision is higher than the probability of the other processes shown in Figure 45. This is so because gluons have, to put it crudely, double color charges (they are red-antiyellow, blue-antired, and so on). By studying the production of gluon jets, we can obtain detailed information about the gluon-gluon interaction. This is a fascinating and important problem. In particular, it would be very interesting to verify the reality of the four-gluon interaction (Figure 45f) predicted by quantum chromodynamics.

I have not mentioned the search for the t quark in the overview of collider physics only because I am sure that by now the reader himself is anxiously awaiting its discovery.[⊊]

"Quiet physics" and grand unification

It would be wrong to think that all important future experiments in high-energy physics will inevitably involve giant accelerators. Actually, there has been considerable interest recently in "quiet physics", i.e., in experiments conducted deep underground, in the so-called low-background laboratories.

The first to note are experiments on the stability of the proton. We know that protons are extremely stable: their lifetime exceeds the age of the Earth by at least 20 orders of magnitude. The reason it is possible to find such a high lower limit for the proton lifetime is that the protons around us are so abundant: one gram of hydrogen contains $6 \cdot 10^{23}$ protons. Consequently, if we take, say, 9 tons of water (i.e., 1 ton of free protons) and show that not a single proton decays in a year, we can conclude that the lifetime of the free proton is greater than $6 \cdot 10^{29}$ years. The most recent and most accurate experiments of this kind made use of a detector containing 8000 tons of water. The huge detector was built in a deep mine in order to reduce the cosmic-ray background. The experimentalists were looking for $p \rightarrow e^{+}\pi^{0}$ decays and concluded that the proton lifetime in this decay channel was greater than $1.9 \cdot 10^{31}$ years. When protons in oxygen nuclei are also taken into account, this figure goes up to $6.5 \cdot 10^{31}$ years.

Proton decay attracted considerable attention about ten years ago in

[⊊]See footnote on p. 33.

connection with the advent of *grand unification theories*. These theories or, more precisely, theoretical models are attempts at a unification of the strong, weak, and electromagnetic forces in which all these interactions are described by a single common charge g. The question is: is such a theory possible in principle if experiments show that $g_S > g_W > e$?

The key to this problem is a phenomenon called *vacuum polarization*. This phenomenon resembles the polarization of a dielectric into which an electric charge has been inserted. The electric charge of the electron is partly neutralized or screened by virtual electron-positron pairs or by pairs of other particles and antiparticles in the physical vacuum. This occurs because the electron attracts positive and repels negative particles so that, when its charge is observed from a large distance, it appears to be smaller than its "bare" charge. The deeper we probe into the cloud of pairs surrounding the electron, the greater is the apparent electric charge of the electron. The electric charge of any particle increases as the distance to the particle decreases or, by virtue of the uncertainty relation, as the momentum q transferred to the particle increases. The electric charge e(q) is thus an increasing function of q. Similarly, the strong and weak charges are also functions of q.

Similarly, the strong and weak charges are also functions of q:

$$g_S = g_S(q) \text{ and } g_W = g_W(q).$$

However, in contrast to the electric charge, $g_S(q)$ and $g_W(q)$ are found to decrease with increasing q. This striking property of the weak and strong charges is called the *asymptotic freedom* (for asymptotically large momentum, the charges g_S and g_W tend to zero, the weak and strong interactions are "turned off", and the particles behave as free noninteracting particles).

The manifestation of asymptotic freedom is strongest for the strong charge g_S. It is caused by the specific interaction between gluons and constitutes the "reverse side" of confinement. For example, for q = 1 GeV, we find that $\alpha_S \simeq 0.3$; for q = 100 GeV, the result is $\alpha_S \simeq 0.1$. Thereafter α_S continues to fall with increasing q (we recall that $\alpha_S = g_S^2/\hbar c$, $\alpha_W = g_W^2/\hbar c$, and $\alpha = e^2/\hbar c$).

If we extrapolate this variation of the "constants" $\alpha_S(q)$ and $\alpha_W(q)$ to large values of q, we find that they both become equal to approximately 1/40 for $q \simeq 10^{14} - 10^{15}$ GeV. It is striking that the electromagnetic "constant" $\alpha_{EM}(q)$ is also equal to 1/40 at this point [$\alpha_{EM}(q) = (8/3)\alpha(q)$, where the coefficient 8/3 is furnished by the theoretical models of grand unification: at the grand unification point, $\sin^2\theta_W = 3/8$]. Thus, at $q \simeq 10^{15}$ GeV,

$$\alpha_{EM} = \alpha_W = \alpha_S = \alpha_{GU} \simeq 1/40.$$

(the subscript GU stands for Grand Unification).

Figure 46 illustrates the meeting of the three "running constants". A glance at this drawing (a map of the three fundamental forces) reminds us of the famous α-β-γ drawing on p. vi, which symbolizes our first encounter with these three forces. We have come a long way since then.

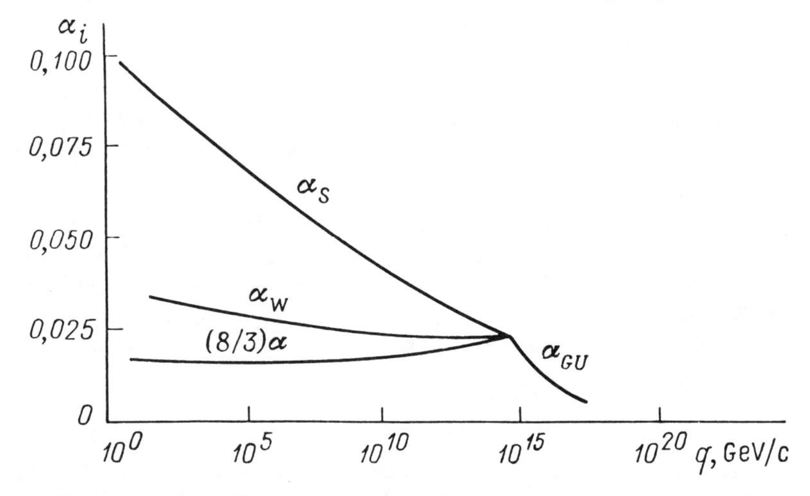

Figure 46. The effective charges of the strong (g_S), weak (g_W), and electromagnetic (e) interactions as functions of the momentum transferred to a particle.

$\alpha_S = g_S^2/\hbar c$, $\alpha_W = g_W^2/\hbar c$, $\alpha = e^2/\hbar c$, $\alpha_{GU} = g_{GU}^2/\hbar c$

where g_{GU} is the charge of the unified, universal interaction.

According to the simplest model of grand unification, color and electroweak symmetries merge at $q \simeq 10^{15}$ GeV into a unified symmetry comprising 24 vector bosons. Half of them are the familiar gluons, intermediate bosons, and the photon. The other half consists of super-superheavy particles, the so-called X and Y bosons with masses of about 10^{15} GeV. Like quarks, each of these bosons has three colors, so that there are 12 of them (including the corresponding antiparticles $\tilde{X}$ and $\tilde{Y}$). X bosons have an electric charge of $\frac{4}{3}$ and Y bosons have a charge of $\frac{1}{3}$.

The striking property of the X and Y bosons is that both interact with two different kinds of current, namely, two-quark and quark-lepton

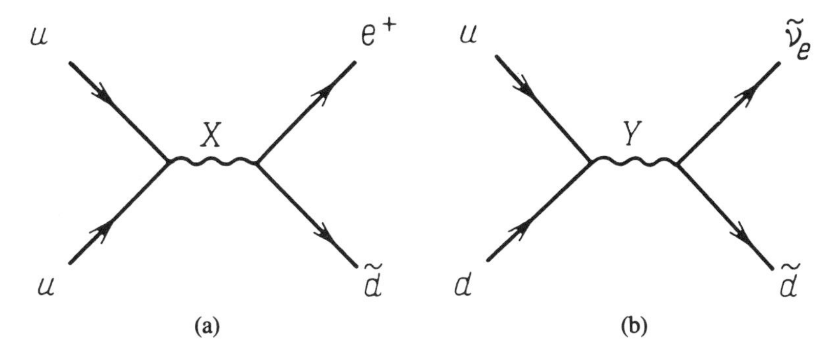

Figure 47. Processes involving the X and Y bosons and resulting in the transformations of a proton into a positron (a), and a neutron into an electron neutrino (b).

currents (Figure 47). If we replace the outgoing $\tilde{d}$ quark in Figures 47a and 47b with an ingoing d quark, we can readily see that the former describes the transformation of a proton into a positron, and the latter describes the transformation of a neutron into an antineutrino. In practice, this transformation must be accompanied by the emission of an additional meson (π^0) or of several mesons, in order to satisfy the momentum and energy conservation laws.

If we use Figure 47 to evaluate the proton lifetime, τ_p, we find that

$$\frac{1}{\tau_p} \simeq \frac{\alpha_{GU}^2 m_p^5 c^2}{m_x^4 \hbar},$$

where $\alpha_{GU} \simeq 1/40$ is the grand unification constant, m_X is the X boson mass, and m_p is the proton mass (compare this formula with that for the muon lifetime). If we convert this result into the familiar time units, we readily find (the reader should try this) that the expected proton lifetime is about 10^{30} years. More rigorous calculations confirm this estimate: they show that the simplest model of grand unification does indeed predict that the proton lifetime is not greater than 10^{30} years, which conflicts with the experimental lower limit of close to 10^{32} years. More sophisticated versions of grand unification, which do not contradict the experiment, have been developed. Unfortunately, we still lack the foundation for a judicious choice between these versions.

It would be unfortunate if the proton were to live for, say, 10^{35} years. Indeed, it would take a few decades to build large enough and sophisticated enough detectors capable of recording such slow decays.

Even if the diameter of our accelerators were to reach several times the diameter of the Earth, we would not be able to approach the grand unification energy or the threshold for the production of X and Y bosons. This will be clear if we take the energy of the Tevatron (1 TeV) and multiply it by the ratio of the Earth's equator (40 000 km) to the circumference of the Tevatron (7 km). All we can do, therefore, is to hope that nature is compassionate. Actually, there are just a few other phenomena that could, in principle, throw some light on how nature manages to achieve grand unification.

One of these is the transformation of neutrons into antineutrons in vacuum, with a period of a few years or more. As these lines are being written, several experimental groups are searching for these neutron-antineutron vacuum oscillations in neutron beams generated by nuclear reactors.

The problem of neutrino mass and of the mutual transformations of different types of neutrino in vacuum, namely,

$$\nu_e \to \nu_\mu,\ \nu_\tau;\ \nu_\mu \to \nu_e,\ \nu_\tau;\ \nu_\tau \to \nu_e,\ \nu_\mu$$

has recently attracted particular attention, mostly within the framework of the grand unification problem.

In fact, we want to find out whether the triple of neutrinos (ν_e, ν_μ, ν_τ) is "rotated" relative to the triple of charged leptons (e, μ, τ) (cf. Figure 28), by analogy with the relative rotation of the triples of "upper" and "lower" quarks in Figure 29. The subtle point here is that, if the neutrino mass is zero, this rotation is unobservable in principle. Note that, if the masses of the u, c, and t quarks (or d, s, and b quarks) were zero, or even if they were merely equal to one another, the relative rotation of the two quark triples would be equally unobservable.

In order to clarify the matter, consider Figures 48 and 49. From what has been said above about the weak charged current, it follows that we can write it in the form

$$\bar{\nu}_e' e + \bar{\nu}_\mu' \mu + \bar{\nu}_\tau' \tau + \bar{u}' d + \bar{c}' s + \bar{t}' b$$

where primed states are linear combinations of unprimed states (see Figures 48c and 49c). If the masses of the three unprimed states were all zero (or simply equal to one another), all physically observable phenomena would involve only the primed states, so we would never be able to deduce the existence of unprimed states.

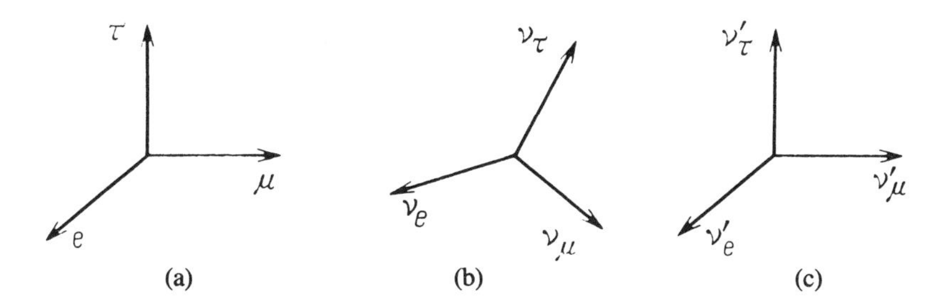

Figure 48. Three mutually orthogonal unit vectors representing three charged leptons (a) and three neutrinos (b). Primed neutrino unit vectors (c) are obtained by rotating the unprimed unit vectors, and are parallel to the unit vectors of the corresponding charged leptons.

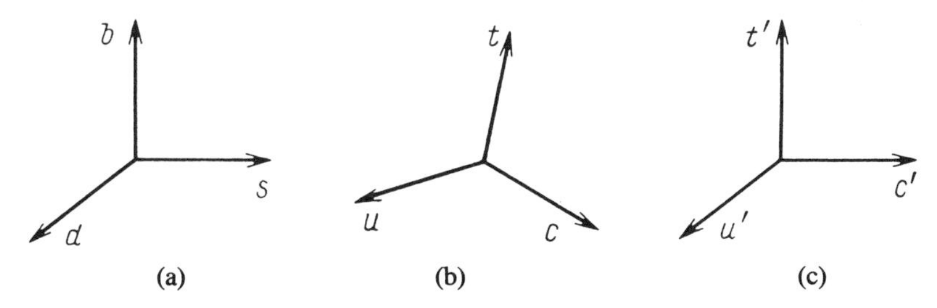

Figure 49. Three mutually orthogonal unit vectors corresponding to the three "lower" quarks (a) and the three "upper" quarks (b). Primed unit vectors of the upper quarks (c) are obtained by rotating unprimed unit vectors, and are parallel to the unit vectors of the corresponding lower quarks.

Since the masses of unprimed states are nonzero and unequal, the wave function describing free propagation in vacuum for each of them has the factor $e^{ipr/\hbar}$ with the characteristic momentum

$$|\mathbf{p}| = \sqrt{(E^2c^{-2} + m^2c^2)},$$

which produces the mutual transformations of primed states in vacuum. Therefore, the vacuum neutrino oscillations will occur if the neutrino triple is rotated as shown in Figure 48 and, at the same time, the masses of at least two neutrino flavors are unequal.

So far, the search for neutrino oscillations in neutrino beams from nuclear reactors and accelerators has failed to reveal this effect. The negative result could be due to the rotation angles being very small, or the mass differences being very small, or both. If we assume that rotation angles (Figure 48) are not small, experimental data indicate that the difference between squares of the neutrino masses is substantially less than 1 eV^2.

Grand unification models do not provide unambiguous answers to questions about neutrino masses. However, the majority of these models predicts very low masses (less than, or about, 10^{-5} eV).

To conclude the subject of neutrinos, we note the interesting possibility that ν_e, ν_μ, and ν_τ are truly neutral particles, each being as ''single'' as the photon. If this is so, the neutrino and antineutrino are actually two polarization states of the *same* particle. These states differ by the mutual orientation of spin and momentum. Of course, leptonic charge is then not a good quantum number. Thus, if the neutral electron neutrino has a nonzero mass, there must be processes violating leptonic charge conservation, for instance, the neutrinoless double β-decay of atomic nuclei. The neutrinoless double β-decay is caused by the simultaneous decay of two neutrons exchanging a virtual neutrino:

$$n + n \rightarrow p + p + e^- + e^-.$$

An example of this decay, which is being looked for in a number of laboratories, is $^{48}Ca_{20} \rightarrow {}^{48}Ti_{22} + e^- + e^-$.

Another interesting object that appears in grand unification models (at least, on paper, at the tip of a theorist's pen) is the magnetic monopole, i.e., an isolated magnetic charge. According to the theory, the mass of this object must exceed by about two orders of magnitude even the mass of the X and Y bosons. Clearly, such heavy objects can never be produced in laboratories. We can only hope that a monopole coming from the depths of the Universe will fly into an experimental setup awaiting this surviving witness of the early times, when the Universe was so hot that pairs of monopoles could be created, but was expanding too fast for these pairs to annihilate.

We shall discuss the hot Universe later. For the moment, we turn to the prospects for superunification, i.e., for a unified theory of electromagnetic, weak, strong, and gravitational forces.

Superunification?

At the beginning of this book, we mentioned a satellite revolving around the Earth. We have used this as a springboard for a discussion of the binding energy of an electron, but we never returned to the gravitational interaction. This is partly justified by the fact that, under laboratory conditions, the gravitational interaction is extremely weak. The potential energy of gravitational attraction between two slow protons is

$$U(r) = -G_N m_p^2/r,$$

where r is the distance between the protons, m_p is the proton mass, and G_N is Newton's constant. In SI units,

$$G_N = 6.7 \cdot 10^{-11}\ m^3kg^{-1}s^{-2}.$$

It is difficult to see, when these units are used, whether the constant G_N is large or small, but if we convert it into electron volts and, as always, take out ħ and c, we see that Newton's constant is exceedingly small:

$$G_N = 6.7 \cdot 10^{-39} \hbar c^5\ GeV^{-2}.$$

In terms of natural units, $\hbar = 1$, $c = 1$, the constant G_N is less by approximately 33 orders of magnitude than the constant of the weak four-fermion interaction (the reader should check this):

$$G_F = 1.2 \cdot 10^{-5} \hbar^3 c^3\ GeV^{-2}.$$

The gravitational attraction between two slow protons is weaker by approximately 36 orders of magnitude than their Coulomb repulsion. For two electrons, the ratio exceeds 10^{42}. However, as the energy E of the colliding electrons increases, the gravitational interaction between them grows stronger. The crucial fact is that the potential of this interaction increases as E^2, being proportional to m^2 only for slow nonrelativistic particles. It is readily verified that, for $E \approx 10^{18}$ GeV, the gravitational attraction between two colliding electrons overcomes their Coulomb repulsion, and if we take $E \approx 10^{19}$ GeV and an impact parameter $r \approx \hbar c/10^{19}\ GeV \approx 10^{-33}$ cm, the gravitational potential energy becomes equal to the kinetic energy of electrons; the gravitational interaction becomes really strong.

The quantity

$$m_{Pl} = (\hbar c/G_N)^{\frac{1}{2}} \simeq 1.2 \cdot 10^{19}\ GeV/c^2 \simeq 10^{-5}\ g$$

is called the Planck mass. At energies and momentum transfers of the order of m_{Pl}, the gravitational interaction becomes dominant.

We note that the Planck mass differs from the grand unification energy by only 4–5 orders of magnitude and by an even smaller factor as compared with the monopole mass. This entitles us to expect that the fundamental laws of the physical world are forged in the vicinity of the Planck mass.

According to modern theoretical concepts, the quanta of the gravitational field are the *gravitons*, i.e. massless spin-2 fundamental bosons. The fact that the graviton mass (like the photon mass) is zero corresponds to the slow ($1/r$) decrease of the Newtonian potential (like that of the Coulomb potential) at large distances.

The fact that the graviton is a spin-2 particle is responsible for the important differences between gravitational and electromagnetic interactions (we recall that the photon is a spin-1 particle). One of the main differences is that the exchange of gravitons between any two particles invariably produces an attraction between them, while the exchange of photons produces attraction between particles of unlike electric charge and repulsion between particles of like charge. Ordinary matter being electrically neutral (electron charges screen the charges of nuclei), the electric forces between atoms are small for distances much greater than the linear dimensions of the atom. There is no such mutual screening of "gravitational charges" in the case of gravitational attraction: all these "charges" are of "like sign". This is why the laws governing the motion of heavenly bodies are governed by gravitation.

■ All this happens because gravitons are tensor particles: they have spin 2. Being tensor rather than vector particles, they are emitted and absorbed not by a vector but by a tensor current, i.e., the energy-momentum tensor. This is not the place for a detailed explanation of what the energy-momentum tensor is and why the exchange of tensor bosons produces an interaction that grows quadratically with increasing energy E of colliding particles. We merely note that the Feynman vertex for the emission and absorption of a tensor particle contains an additional multiplier E if we compare it with a similar vertex for the emission and absorption of a vector particle.■

The gravitational constant is so small and the gravitational interaction at low energies so weak that gravitons could only be detected in the unlikely event of the heavens sending down to an underground laboratory a fantastically high flux of gravitons of fantastically high energy. The quanta of the gravitational field generated under terrestrial conditions are virtually undetectable. The gravitational interaction at energies of the order of the Planck mass is, in its theoretical aspect, the most complex problem that physicists have ever tried to tackle. High hopes are invested in supersymmetry and its offsprings – supergravity and superstrings. However, the ideas of supergravity and superstrings are much too complicated to be discussed here.

The experimental aspect of the situation looks utterly bleak and hopeless because there can be no doubt that the Planck mass will never be reached with accelerators. The question is how to determine without laboratory tests which of the competing theories is valid and which are wrong? The answer is that, despite the fact that there are no Planck-mass accelerators and never will be, a Planck-mass laboratory does exist. This laboratory is the early Universe.

Cosmology and astrophysics

Astronomical observations lead to a number of conclusions of fundamental significance. Here are the most important.

1. Remote galaxies recede both from us and from one another, the recession velocity v of any two galaxies being the greater, the greater the distance R between them. The proportionality factor H between v and R is called the Hubble constant. This factor changes as the Universe grows older. At present,

$$H \approx 60 \text{ kms/s per megaparsec}$$

or

$$H \approx 18 \text{ km/s per } 10^6 \text{ light years.}$$

It is readily seen that 1/H has the dimensions of time. It can be shown that 1/H is a measure of the age t of the Universe. More precisely, $t = 1/2H \approx 10^{10}$ years.

About 10^{10} years ago, the Universe was very young, very dense, and very hot, having been created in the so-called Big Bang. The theory of the Big Bang establishes the following relation between the age t and temperature T of the Universe:

$$t = m_{Pl}c^2\hbar/(kT)^2,$$

where k is the Boltzmann constant [k is the conversion factor between kelvins (K) and energy units: $1 \text{ K} \approx 10^{-4}$ eV]. The relation between the age of the Universe and its temperature can be written as follows:

$$t(\text{in seconds}) \approx 1/T^2 \text{ (in MeV}^2\text{)}$$

2. In the first few seconds of the Big Bang, the Universe was dominated by the abundance of very energetic gamma rays. As the Universe was expanding and cooling, the wavelengths of these gamma rays were increasing and their energies were decreasing. At present, we observe them as microwave radiation with a temperature of about 3 K. This *primordial radiation*, discovered in 1965, is isotropic and uniform. One cubic meter of interstellar space contains about 4×10^8 such primordial microwave photons. The primordial photons are approximately a billion times more abundant than hydrogen atoms, hydrogen being by far the most abundant element in the Universe (if all stellar material were uniformly spread throughout the Universe, one cubic meter would contain only one hydrogen atom).

3. In addition to matter in luminous stars and to primordial photons, the Universe contains the so-called dark matter whose total mass is an order of magnitude greater than the mass of visible matter. The presence of the dark matter is revealed by its gravitational effect on the motion of stars and galaxies.

So far, the physical nature of the dark matter remains a mystery. According to one hypothesis, it consists of primordial neutrinos. According to the hot Universe theory, the density of primordial neutrinos must be roughly the same as that of primordial photons. If we accept this, we can give the following estimates: for the total mass of all neutrinos to exceed by one order of magnitude the total mass of all hydrogen atoms, the mass of one neutrino must be about 20–30 eV (the reader should verify this).

Now that we have touched upon the subject of neutrinos in the Universe, we have to mention other aspects of this topic. For example, the abundance of helium in the Universe is very sensitive to the number of neutrino flavors. The observed helium abundance agrees best with the hypothesis that no flavors of the neutrinos other than the ν_e, ν_μ, and ν_τ are allowed in nature.

It was mentioned at the beginning of this book that neutrinos are emitted in fusion reactions within stars. A special underground experiment, designed to detect neutrinos arriving from the central regions of the Sun, has demonstrated that the flux of solar neutrinos with energy higher than a few MeV is only about one-third of what theoretical astrophysicists have predicted. A possible explanation is that the model of the solar interior on which the calculations are based has to be refined. Another possibility is that the neutrino oscillations convert the flux of electron neutrinos leaving the Sun and heading for the Earth into a mixture of ν_e, ν_μ, and ν_τ. The detector would not then notice ν_μ and ν_τ neutrinos with energies below the muon and tau lepton production thresholds.

Let us now leave the Sun and return to the first few moments in the life of the Universe, when its temperature was near the Planck value ($T \simeq m_{Pl}c^2/k \simeq 10^{19}$ GeV/k $\simeq 10^{32}$ K) or close to the grand unification temperature ($T \simeq 10^{28}$ K). It was during these first moments (at $t \simeq 10^{-43} - 10^{-35}$ s) that the main characteristics of our Universe, including its density and the ratio of the number of protons to that of photons, were "preprogrammed". The participants in this "preprogramming" were not only the lighter fermions and bosons that we produce and study in accelerators but also the superheavy giants, like the X and Y bosons.

It appears more and more plausible that Planck-scale physics determines not only all of the physics at lower energies but also the entire organization of our Universe.

In praise of high-energy physics

Each year, tens and hundreds of millions of rubles, dollars, francs, and marks, and tens and hundreds of billions of lira and yen are spent on particle physics. Is this expenditure justified? One of our aims in this book is to show that this expenditure is more than justified.

In a certain sense, the money spent on high-energy physics is like the money spent on our children: neither is the best investment for immediate financial return. Nevertheless, the world is unthinkable without children, and the future of science is unthinkable without particle physics. We shall presently return to the basic role of particle physics, but first let us look at some examples of spin-offs that high-energy physics has already produced.

We can begin with the diverse and constantly proliferating applications of particle beams generated by accelerators. For example, proton beams are used in medicine to fight tumors. Unlike X-rays and gamma rays, protons can be sent out as beams of very small cross section. They deposit most of their energy at the end of their track, at a prescribed depth, so that the damage to healthy tissue surrounding the tumor is minimized.

Muons are used to study very subtle physicochemical properties of various materials. Because of parity nonconservation, the electron appearing as a result of a muon decay moves in the direction of the spin of the muon. The magnetic moment of the muon also points along its spin. Consequently, the outgoing electron "informs" experimenters about the orientation of muon magnetic moment. A muon thus functions as a tiny "talking magnetic needle" that measures the distribution of magnetic fields in materials, for example, in crystals. A positively charged muon and an ordinary electron form the hydrogen-like muonium atom (μ^+e^-), which has the same size as the hydrogen atom but is about ten times lighter. The chemical properties of muonium are close to those of hydrogen, but the fate of a muonium atom is much easier to trace.

Another promising practical application of muons is the muon catalysis of nuclear fusion reactions. The size of an atom of muonic hydrogen [μ^-p, μ^-D, or μ^-T atoms (D for deuteron and T for triton) in which a muon plays the role of the electron] is approximately 200 times less (in the ratio of m_μ to m_e) than the size of the ordinary, electronic, hydrogen atoms (e^-p, e^-D, or e^-T). This means that a neutral atom of muonic hydrogen can approach very closely the nucleus of the ordinary electronic hydrogen atom. Consequently, two nuclei of heavy hydrogen (e.g., D and T) can be brought together by overcoming their Coulomb repulsion, not by way of high temperture, but by using the "muon cover". The muon thus plays the role of a Trojan horse: its negative charge screens the positive charge of heavy hydrogen, so the fusion reaction $T + D \rightarrow He^4 + n$ can proceed at low temperatures, in liquid hydrogen.

Muons generated by cosmic rays and traversing the Great Pyramids have been used as highly penetrating "rays" to study the inner structure of these monuments. It has also been proposed to use high-energy muon beams for "x-raying" the Earth's crust and for locating ore deposits.

Synchrotron radiation emitted by high-energy electrons moving along their orbits in a circular accelerator is widely used in science, medicine, industry, and agriculture. Applications range from the structural analysis of the living cell to producing a moving picture of a patient's heart, or to the extermination of pests in large volumes of grain.

By detecting antineutrinos emitted in nuclear reactors, it is possible to monitor the intensity of fission processes and, hence, the consumption of nuclear fuel better than by any other method. This kind of "neutrino meter" is already being used at one of the nuclear power plants.

Recent suggestions include the use of high-energy neutrino beams for geological prospecting (prospecting for oil) and for precise measurements of continental drift.

Neutrino astronomy offers a unique opportunity for detecting the collapse of a star or for examining the solar interior. Indeed, a photon diffuses from the center of the Sun to its surface for about a hundred thousand years, while a neutrino escapes practically instantaneously. Sufficiently powerful neutrino telescopes will be able to monitor processes occurring inside the Sun and produce long-term forecasts.

High hopes are pinned on heavy-ion beams that can trigger nuclear fusion reactions in special pellets composed of isotopes of light elements. Such heavy-ion fusion reactors may become the basis for power engineering in the 21st century.

When speaking about high-energy physics, we must not forget its numerous scientific and technological innovations that have subsequently been widely used in other fields of physics. Because of its exceptionally high requirements for accuracy, response time, energy, and intensity, high-energy physics demands exceptional performance from the instruments and systems that it uses. For example, high-energy physics was the cradle of extremely efficient vacuum pumps and the instigator of the mass production of superconducting magnets. Different, extremely sensitive detectors of ionizing particles were invented for use in high-energy physics, and fast electronics came of age in this field.

However, the true value of elementary-particle physics does not lie in any or all of these numerous applications. This science is the bedrock of all natural science, and this fact determines its exceptional and outstanding role. The level attained in this science determines our level of understanding of the world that surrounds us, and serves as a measure of the intellectual maturity of mankind. Without elementary-particle physics, it would be impossible to reconstruct the past of our world, to understand the main processes taking place in it, and to predict its future.

The lesson that we draw from the history of physics is that each successful step toward the comprehension of the fundamental laws of nature invariably produces enormous (and nearly always unexpected) changes in technology and drastically affects the life of mankind as a whole. Suffice it to recall the benefits brought about by such abstract theories as electrodynamics, relativity, and quantum mechanics. There is no reason to expect that the theories we are elaborating today will be less fruitful. Indeed, they deal with the most fundamental concepts of the physical world, such as time, space, vacuum, mass, charge, and spin. Future studies will uncover totally unexpected aspects of these fundamental entities. There can hardly be any doubt that the practical consequences of these future discoveries will be of fundamental importance, although, at present, we can only guess at what this could be.

Isacc Newton once compared himself to a boy playing with pebbles on the beach of the Ocean of the Unknown: ". . the great ocean of truth lay all undiscovered before me." Surely, the region of the Known has grown enormously since then, but it would be naive arrogance to think that we know nearly all there is to know and that only minor details will have to be scrutinized. Just imagine how much we would miss if the development of physics stopped, say, in 1950 or even as late as 1970. Indeed, *eight* of the sixteen known fundamental fermions and bosons were discovered after 1970 (the c, b, and t (?) quarks, the tau lepton and tau neutrino; the gluon and the W and Z bosons).

As in Newton's time, the great Ocean of the Unknown still lies before us. New particles, new forces, new fundamental principles are waiting to be discovered.

Recommended Literature

1. R. Feynman *The Character of Physical Law*, London, 1965.
2. S. Weinberg *The First Three Minutes: A Modern View of the Origin of the Universe* Basic Books, Inc., New York, 1977.
3. H. Fritzsch *Quarks: Urstoff unserer Welt* P. Piper & Co. Verlag, München, Zürich, 1981 (in German).
4. H. Weyl *Symmetry* Princeton University Press, 1952, Princeton, New Jersey.
5. *The Nature of Matter* Wolfson College Lectures, 1980. Ed. By J H Mulvey, Clarendon Press, Oxford, 1981 (Lectures by D Wilkinson, R Peierls, C Llewellyn Smith, D Perkins, A Salam, J Ellis, J Adams, and M Gell-Mann).
6. L. Okun *Particle Physics: The Quest for the Substance of Substance* Harwood Academic Publishers, New York, 1985.
7. Y. Nambu *Quarks: Frontiers in Elementary Particle Physics* World Scientific, 1985.

If your destiny is physics, some day something may remind you of this book, dedicated to the memory of Isaak Pomeranchuk, and you may wish to find out more about his work.

His papers were published in Russian in three volumes:

I.Ya. Pomeranchuk *Collected Works: vol. 1, Low-Temperature Physics, Neutron Physics; vol. 2, Particle Physics, Electromagnetic and Weak Interactions; vol. 3, Particle Physics, Strong Interactions* Nauka, Moscow, 1972.

Subject Index